Careers in Engineering

Careers in Engineering

SALEM PRESS
A Division of EBSCO Information Services, Inc.
Ipswich, Massachusetts

GREY HOUSE PUBLISHING

Cover photo: iStock

Careers in Engineering, published by Grey House Publishing, Inc., Amenia, NY, under exclusive license from EBSCO Information Services, Inc.

∞ The paper used in these volumes conforms to the American National Standard for Permanence of Paper for Printed Library Materials, Z39.48 1992 (R2009).

Publisher's Cataloging-In-Publication Data
(Prepared by The Donohue Group, Inc.)

Title: Careers in engineering.
Other Titles: Careers in—
Description: Ipswich, Massachusetts : Salem Press, a division of EBSCO Information Services, Inc. ; Amenia, NY : Grey House Publishing, [2022] | Includes bibliographical references and index.
Identifiers: ISBN 9781637000939 (print) | ISBN 9781637000946 (ebook)
Subjects: LCSH: Engineering—Vocational guidance—United States.
Classification: LCC TA157 .C37 2022 (print) | LCC TA157 (ebook)| DDC 620.0023—dc23

First Printing

PRINTED IN THE UNITED STATES OF AMERICA

CONTENTS

PUBLISHER'S NOTE

The latest volume in our *Careers in* series, *Careers in Engineering*, provides a thorough introduction to the wide range of career opportunities available to those interested in the field of engineering.

The types of careers in engineering are incredibly wide-ranging, from the obvious, like civil engineers and environmental engineers, to the not-so-obvious, like building inspectors, drafters, and even sales positions. *Careers in Engineering* offers insight into what it's like to work for an engineering firm, conduct fieldwork, and create and sell cutting-edge devices. If you are interested in how to turn your passion for technology, mechanics, and inventing into a fulfilling career, this book is for you.

At the heart of this volume, like the dozens of previous *Careers in* titles, are 28 occupation profiles that cover various careers that are involved in engineering in some way, from mathematician to medical scientist, to postsecondary teacher. Each career profile offers details about its specific career:

Snapshots offer current data on:

- Median pay
- Job outlook
- Educational and other requirements
- Working conditions
- Recommended areas of interest

Career Overviews describe:

- Duties
- Work environment
- Work schedules

In addition, each profile provides important qualities for the job, specific and general job requirements, anticipated rates of growth, and why. Also included is a list of related occupations, both in the public and private sectors, and a list of resources for more information, with profession-specific associations and certifying bodies.

Merging scholarship with occupational development, this single comprehensive guide-book provides students who are interested in exploring a career in engineering with the necessary insight into the wide array of options available in this popular field. The book offers guidance regarding what job-seekers can expect in terms of training, advancement, earnings, job prospects, working conditions, relevant associations, and more. Careers in Biology is specifically designed for a high school and undergraduate audience and is edited to align with secondary or high school curriculum standards.

Additional Materials

Careers in Engineering is enhanced with statistical data, including projections from the U.S. Bureau of Labor Statistics, and median annual salaries or wages for those occupations profiled. Fast Facts, Famous Firsts, and dozens of photos add depth to the discussion.

A popular and valuable highlight of the *Careers in* series are "Conversations With"—interviews with real-life professionals working in relevant jobs. Those interviewed offer their personal career paths, detail the potential for career advancement, and offer advice for students.

The back matter in this volume includes:

Appendix A: Guide to the Holland Code—John Holland's theory that people and work environments can be classified into six different groups: realistic, investigative, artistic, social, enterprising, and conventional. These interest codes are found at the beginning of each profile.

Appendix B: General Bibliography—a list of suggested readings for those who wish to learn more about pursuing a career in the music industry.

Appendix C: List of Organizations and Resources—a list of additional resources for education, certification, professional development, and networking.

This volume also includes a comprehensive subject index.

Salem Press thanks the professionals whose honest responses and personal insight provide immeasurable value to *Careers in Engineering*.

INTRODUCTION

Engineering is an incredibly varied field of work that affects peoples' lives in ways that many might not consider, and is far too vast to cover in a single volume such as this. Engineering impacts nearly every aspect of the modern world, from the buildings in which we reside to the cars we drive, to the computers that power our daily routines, and the technology that protects our precious environment. The list of careers that are related to engineering is long, including everything from Aerospace Engineer to Sales Engineer.

In broad terms, engineering is the study of how complex objects—such as machines and structures—are built, and how they function. The discipline is an intersection of math and science, with a strong dose of creativity, ingenuity, and artistry.

Because of the profession's varied applications, an individual is rarely just an "engineer," but rather falls into one or more sub-fields of expertise. Anyone interested in flight, agriculture, computers, construction, health and safety, or marine matters can find a career path in engineering. Other scientific disciplines such as physics are closely related.

Finding the path that's right for you means thinking about your own interests, skills, and talents. Important skills for working in engineering include analysis, communication, critical-thinking, interpersonal, math, and problem-solving, not to mention a love of science and a desire to improve the world around us.

It's also worth thinking about how much money you want to make. A position that requires a four-year degree will earn you more money than one requiring a two-year degree, or one requiring no post-secondary education at all. The profiles in this volume include education requirements and typical earnings. Remember, however, that how much money you make is not the entire story. Benefits, job security, where you work, and self-fulfillment are important factors as well.

The "Conversations With..." spread throughout this volume show the variety of career paths available in the field of engineering, how to achieve them, and what to expect when you get there. They are interviews with real individuals working in the field at real jobs.

The list of jobs contained in this volume is not exhaustive, and should be viewed as an entry point into a world with numerous branches and sub-disciplines, many of which can intersect in novel ways. The aim is to provide readers—especially students embarking on their lifelong careers—with accurate and detailed examples of some the many possibilities available under the awning of engineering.

Here are details about how certain careers can relate to engineering:

Construction and building inspectors ensure that construction meets local and national building codes and ordinances, zoning regulations, and contract specifications.

Drafters use software to convert the designs of architects and engineers into technical drawings. Most workers specialize in architectural, civil, electrical, or mechanical drafting and use technical drawings to help design everything from microchips to skyscrapers.

Electro-mechanical and mechatronics technologists and technicians combine knowledge of mechanical technology with knowledge of electrical and electronic circuitry. They operate, test, and maintain unmanned, automated, robotic, or electromechanical equipment.

Marine engineers and **naval architects** design, build, and maintain ships, from aircraft carriers to submarines and from sailboats to tankers. Marine engineers are responsible for the internal systems of a ship, such as the propulsion, electrical, refrigeration, and steering systems. Naval architects are responsible for the ship design, including the form, structure, and stability of hulls.

Postsecondary teachers instruct students in a variety of academic subjects beyond the high school level—in this case, engineering. They may also conduct research and publish scholarly papers and books, while working in the field.

Sales engineers sell complex scientific and technological products or services to businesses. They must have extensive knowledge of the products' parts and functions and must understand the scientific processes that make these products work.

Aerospace Engineer

Snapshot

Career Cluster(s): Science, Technology, Engineering & Mathematics

Interests: Science; Technology; Aviation; Space; Mathematics; Engineering

Earnings (Yearly Average): $118,610

Employment & Outlook: As Fast As Average

OVERVIEW

Sphere of Work

Aerospace engineers design primarily aircraft, spacecraft, satellites, and missiles. In addition, they create and test prototypes to make sure that they function according to design.

Work Environment

Aerospace engineers typically work in an office environment, although they may be required to go into the field to oversee or review the progress of projects they designed.

Occupation Interest

Photo via iStock/3DSculptor. [Used under license.]

People interested in becoming an aerospace engineer are problem-solvers by nature, with a love of facts and figures. Science and mathematics-oriented, they possess a knack for fine technical details. A science, technology, engineering, and mathematics (STEM) education path is a virtual necessity. This occupation while it can provide many hours of solitary work, also requires a person to have strong communication skills as they will be frequently in contact with others and need to convey their complex ideas effectively.

Profile

Working Conditions: Inside
Physical Strength: Light to Heavy Work (Varies)
Education Needs: Bachelor's Degree
Licensure/Certification: Optional
Opportunities for Experience: Cooperative Programs; Internships
Interest Score: IR

A Day in the Life—Duties and Responsibilities

Aerospace engineers may develop new technologies for use in aviation, defense systems, and spacecraft. They often specialize in areas such as aerodynamic fluid flow; structural design; guidance, navigation, and control; instrumentation and communication; robotics; and propulsion and combustion.

Duties and Responsibilities

- Directing and coordinating the design, manufacture, and testing of aircraft and aerospace products
- Assessing proposals for projects to determine if they are technically and financially feasible
- Determining if proposed projects will result in safe operations that meet the defined goals
- Evaluating designs to see that the products meet engineering principles, customer requirements, and environmental regulations
- Developing acceptance criteria for design methods, quality standards, sustainment after delivery, and completion dates
- Ensuring that projects meet quality standards
- Inspect malfunctioning or damaged products to identify sources of problems and possible solutions

Aerospace engineers can specialize in designing different types of aerospace products, such as commercial and military airplanes and helicopters; remotely piloted aircraft and rotorcraft; spacecraft, including launch vehicles and satellites; and military missiles and rockets.

Aerospace engineers often become experts in one or more related fields: aerodynamics, thermodynamics, materials, celestial mechanics, flight mechanics, propulsion, acoustics, and guidance and control systems.

Aerospace engineers typically specialize in one of two types of engineering: aeronautical or astronautical.

OCCUPATION SPECIALTIES

Aeronautical Engineer

Aeronautical engineers work with aircraft. They are involved primarily in designing aircraft and propulsion systems and in studying the aerodynamic performance of aircraft and construction materials. They work with the theory, technology, and practice of flight within the Earth's atmosphere.

Astronautical Engineer

Astronautical engineers work with the science and technology of spacecraft and how they perform inside and outside the Earth's atmosphere. This includes work on small satellites such as cubesats, and traditional large satellites.

WORK ENVIRONMENT

Immediate Physical Environment

Aerospace engineers are employed in industries in which workers design or build aircraft, missiles, systems for national defense, or spacecraft. They work primarily for firms that engage in manufacturing, analysis and design, research and development (R&D), and for the federal government.

Aerospace engineers now spend more of their time in an office environment than they have in the past, because modern aircraft design requires the use of sophisticated computer equipment and software design tools, modeling, and simulations for tests, evaluation, and training.

Aerospace engineers work with other professionals involved in designing and building aircraft, spacecraft, and their components. Therefore, they must be able to communicate well, divide work into manageable tasks, and work with others toward a common goal.

Human Environment

Aerospace engineers work with other professionals involved in designing and building aircraft, spacecraft, and their components. Therefore, they must be able to communicate well, divide work into manageable tasks, and work with others toward a common goal.

Technological Environment

Technological requirements in this position are many, including knowledge of various analytical and scientific software programs (such as MATLAB). Aerospace engineers will use a number of tools, among them computer-aided design (CAD) software programs like computer-aided three-dimensional interactive application (CATIA); physical tools, from mallets to torches to Pneumatic power tools, are frequently employed.

EDUCATION, TRAINING, AND ADVANCEMENT

High School/Secondary

Students interested in aerospace engineering should focus on technical courses such as computer science. They might also consider joining their school's science club, if possible, for practical experience and relationship building.

Suggested High School Subjects

- Algebra
- Geometry
- Physics
- Pre-Calculus
- Trigonometry

Related Career Pathways/Majors

Science, Technology, Engineering & Mathematics Career Cluster

- Engineering & Technology Pathway

Transferable Skills and Abilities

Analytical Skills
- Identifying design elements that may not meet requirements; formulating alternatives to improve the performance of those elements

Business Skills
- Knowing standard business practices/commercial law, and meeting federal government standards

Critical-thinking Skills
- Producing designs that meet governmental standard
- Figuring out why particular designs do not work
- Asking the right questions, then finding acceptable answers

Math Skills
- Using the principles of calculus, trigonometry, and other advanced topics in math for analysis, design, and troubleshooting in their work

Problem-solving Skills
- Upgrading designs and troubleshooting problems when meeting new demands for aircraft, such as increased fuel efficiency or improved safety

Writing Skills
- Writing papers that explain their designs clearly and creating documentation for future reference

Postsecondary

Bachelor's degree programs in Aerospace Engineering include classroom, laboratory, and field studies in subjects such as general engineering principles, propulsion, stability and control, structures, mechanics, and aerodynamics, which is the study of how air interacts with moving objects.

Some colleges and universities offer cooperative programs in partnership with regional businesses, which give students practical experience while they complete their education. Cooperative programs and internships enable students to gain valuable experience and to finance part of their education.

At some universities, a student can enroll in a five-year program that leads to both a bachelor's degree and a master's degree upon completion. A graduate degree will allow an engineer to work as an instructor at a university or to do R&D. Programs in aerospace engineering are accredited by the Accreditation Board for Engineering and Technology, Inc. (ABET).

Fast Fact

The term "aerospace engineering" first appeared in 1958—the same year NASA was founded.
Source: monroeaerospace.com

Photo via iStock/Elen11. [Used under license.]

Related College Majors

- Astronomy
- Electrical Engineering
- Mechanical Engineering
- Nuclear Engineering
- Physics

Adult Job Seekers

Adults interested in becoming an aerospace engineer should have a degree in a related field in order to be familiar with the necessary processes and techniques. They may be required to take additional college courses to supplement their knowledge, and college courses can also provide networking opportunities.

Professional Certification and Licensure

Licensure for aerospace engineers is not as common as it is for other engineering occupations, nor it is required for entry-level positions. A Professional Engineering (PE) license, which allows for higher levels of leadership and independence, can be acquired later in one's career. Licensed engineers are called professional engineers (PEs). A PE can oversee the work of other engineers, sign off on projects, and provide services directly to the public. State licensure generally requires:

- A degree from an ABET-accredited engineering program
- A passing score on the Fundamentals of Engineering (FE) exam
- Relevant work experience, typically at least four years
- A passing score on the Professional Engineering (PE) exam.

The initial FE exam can be taken after earning a bachelor's degree. Engineers who pass this exam are commonly called engineers in training (EITs) or engineer interns (EIs). After meeting work experience requirements, EITs and EIs can take the second exam, called the Principles and Practice of Engineering.

Each state issues its own licenses. Most states recognize licensure from other states, as long as the licensing state's requirements meet or exceed their own licensure requirements. Several states require continuing education for engineers to keep their licenses.

Additional Requirements

Aerospace engineers typically work full time. Engineers who direct projects must often work extra hours to monitor progress, to ensure that designs meet requirements, to determine how to measure aircraft performance, to see that production meets design standards, and to ensure that deadlines are met.

EARNINGS AND ADVANCEMENT

The median annual wage for aerospace engineers was $118,610 in 2020. The lowest 10 percent earned less than $72,770, and the highest 10 percent earned more than $171,220.

Aerospace engineers may qualify for benefits; this is up to the discretion of the employer. They also may be entitled to reimbursement or travel stipends to cover the cost of fieldwork.

EMPLOYMENT AND OUTLOOK

Employment of aerospace engineers is projected to grow 8 percent from 2020 to 2030, about as fast as the average for all occupations.

Aircraft are being redesigned for less noise pollution and better fuel efficiency, particularly with the introduction of electric and hybrid-electric aircraft, which should help sustain demand for aerospace engineers in R&D.

Technological advancements have reduced the cost of launching satellites. As space becomes more accessible, especially with developments in small satellites that have greater commercial viability, demand for aerospace engineers among businesses and government entities is expected to increase. In addition, continued interest in drones will help to drive employment growth for these engineers.

Related Occupations

- Aerospace Engineering/Operations Technician
- Architectural/Engineering Manager
- Computer Hardware Engineer
- Electrical/Electronic Engineering Technologist/Technician
- Electrical/Electronics Engineer
- Industrial Engineer
- Materials Engineer
- Mechanical Engineer

MORE INFORMATION

Accreditation Board for Engineering and Technology, Inc. (ABET)
415 N. Charles Street
Baltimore, MD 21201
410.347.7700
comms@abet.org
www.abet.org

The American Society of Mechanical Engineers (ASME)
2 Park Avenue
New York, NY 10016-5990
800.843.2763
custcare@asme.org
www.asme.org

IEEE Aerospace and Electronic Systems Society
admin@ieee-aess.org
www.ieee-aess.org

International Association of Machinists and Aerospace Workers (IAMAW)
9000 Machinists Place
Upper Marlboro, MD 20772
301.967.4500
www.goiam.org

SpaceTEC
379 Cheney Highway
PMB 207
Titusville, FL 32780
321.567.5193
www.spacetec.us

Technology Student Association (TSA)
1904 Association Drive
Reston, VA 20191-1540
703.860.9000
general@tsaweb.org
www.tsaweb.org

Daniel Vest

Conversation With...

JOHN ROSE

Chief of Staff, Boeing Defense Space and Security
Huntington Beach, CA
Aerospace Engineer, 17 years

What was your individual career path in terms of education/training, entry-level job, or other significant opportunity?

I've always loved aviation and spaceflight, and originally wanted to be an astronaut. By high school, I knew I wanted to be an aerospace engineer and took advanced placement (AP) courses in chemistry and physics. My physics class got to visit the Space Academy in Huntsville, Alabama, which was my second visit there.

I got my BS in aerospace from Cal Poly Pomona. In college, I worked part time on the attractions in the Disneyland Resort and spent a summer interning with Disney's Ride & Show Engineering, Inc., where I later spent two years of my career. I also did two other internships.

As graduation approached, Rockwell International, the builder of the Space Shuttle, asked me to interview. I was thrilled, since working on the Space Shuttle was THE job I wanted. So, my first full-time job in the aerospace industry was in a group supporting the Space Shuttle and International Space Station.

A benefit of working for a large company like Rockwell (which became Boeing) is that they may pay for your advanced degrees. I got my MS in aerospace, also from Cal Poly Pomona, and my MBA from the University of Southern California.

I've always leveraged the opportunities at Boeing. In my current role, in the Office of the Vice President of Engineering, Mission Assurance and Product Support, I'm working with top leadership to develop global strategy on our defense side.

What are the most important skills and/or qualities for someone in your profession?

Aerospace, particularly defense, can be cyclical, so you need to be adaptable. Contracts come and go, sometimes without much notice. Also, new graduates need to recognize that the way something is done in industry may not be how they learned it in class. Inquisitiveness and passion are also important. In addition, the aerospace industry is typically risk averse, so experience is a highly valued commodity. This means younger engineers must be patient; there isn't always a lot of support for putting untested or inexperienced individuals in a critical role.

What do you wish you had known going into this profession?
Aerospace products are a prime target for cybersecurity threats and must be protected from being taken offline or taken control of. Had I known cybersecurity would develop into such an important area, I would have taken related coursework.

Also, I had to learn that it's easy to slip through the cracks at a large company and that it isn't a bad thing to ask about opportunities if they aren't presenting themselves.

Are there many job opportunities in your profession? In what specific areas?
Companies like Boeing or Lockheed Martin offer jobs in a wide variety of areas, including business development, finance, software development, manufacturing, structural design, systems engineering, and computer science.

Unmanned aerial vehicles (UAVs) are fueling a major evolution. Pilotless vehicles can remove human physiological limitations on what a vehicle can do (for example, making sharper turns or diving deeper) as well as spare humans from some dangerous missions. UAVs are adding a new commercial aspect to the industry because they can be made by smaller companies.

How do you see your profession changing in the next five years? What role will technology play in those changes, and what skills will be required?
The fact that most products will be tied to the internet will change how we design and build them, as well as how we approach vulnerabilities relative to such systems as flight controls. Air traffic management is becoming satellite-based, and that's a quickly evolving area of cybersecurity where we need to respond to threats.

Additive manufacturing, or (three-dimensional) 3D printing, is revolutionizing how things are built. Small, portable machines are building pieces used in rocket engines. The concept is also being used on the International Space Station where a 3D computer-aided design and drafting (CADD) model for a specific part or tool can be designed on earth, emailed up to the International Space Station (ISS), and "printed" out.

Immersive development—which is kind of a virtual reality that allows us to try out a repair procedure or do a virtual walkaround of a product without the time and expense of travel—will be a bigger part of the design of and customer support for products.

Systems thinking—being able to conceptualize all of the moving pieces and influencers, versus focusing on one piece—will be valuable.

What do you enjoy most about your job? What do you enjoy least about your job?
I really love the amazing products that our industry produces. I love going to an airshow with my son and seeing them. I also love that many of our products support our armed forces. And I have amazing opportunities to work on projects in different parts of the United States and the world.

But I also have to deal with nontechnical, administrative issues like coordinating meetings that are necessary but not exciting. Another downside is the cyclical nature of being tied to defense and space budgets.

Can you suggest a valuable "try this" for students considering a career in your profession?

Try to find an internship. Also, get to an airshow—seeing all the products and watching them fly is great exposure to the amazing things we build.

This interview was originally published in 2015.

Conversation With...
NARAYANAN M. KOMERATH
Professor of Aerospace Engineering

What was your career path?

I received a Bachelor of Technology in aeronautical engineering, Master of Science and PhD in aerospace engineering, then worked one year as a postdoctoral fellow, and two years as a research engineer, both with a very active experimental research program in the areas of turbulent combustion diagnostics, ramjet combustion acoustics, and rotorcraft aerodynamics. I was then invited to apply for an academic position, becoming assistant professor, associate professor, and eventually promoted to professor in 1994.

What are three pieces of advice you would offer someone interested in your profession?

1. Find something that really appeals to you, not what other people think you like. Essential to stay interested later in life.

2. Work really hard at the beginning—"hit the ground running." It will get you used to working hard, and you will be moving fast enough to catch opportunities.

3. Talk to everyone you can about your field, but do your own thinking to make decisions.

What paths for career advancement are available to you?

- I could apply to become a university administrator (school director, Dean)
- I could apply for positions with the Federal or State government agencies
- could start a company.

However, the advice I receive suggests that the freedom that I have to pursue new interests as a professor is hard to achieve in any other line of work.

This interview was originally published in 2013.

Aerospace Engineering/ Operations Technician

Snapshot

Career Cluster(s): Science, Technology, Engineering & Mathematics

Interests: Science; Technology; Aviation; Space; Mathematics; Engineering

Earnings (Yearly Average): $68,570

Employment & Outlook: Average Growth Expected

OVERVIEW

Sphere of Work

Aerospace engineering and operations technologists and technicians run and maintain equipment used to develop, test, produce, and sustain aircraft and spacecraft.

Photo via iStock/Extreme-Photographer. [Used under license.]

Work Environment

Aerospace engineering and operations technologists and technicians usually work in manufacturing plants, laboratories, and offices. Most work full time.

Occupation Interest

Individuals interested in working as an aerospace engineer/operations technician must love math and science; they typically possess strong analytical skills and problem-solving expertise. Putting their investigative minds to work, they observe fine technical details and solve complex scenarios pertaining to the development of spacecraft. Furthermore, they must be equipped with strong communication skills in order to deliver their complex ideas to others who rely on their expertise. This career is ideal for someone with an interest in the science, technology, engineering, and mathematics (STEM) subjects, especially those with a passion for the spheres of outer space exploration and aviation.

Profile

Working Conditions: Inside
Physical Strength: Light to Heavy Work
Education Needs: High School Diploma; Certificate; Associate's Degree
Licensure/Certification: Optional
Opportunities for Experience: Co-op
Interest Score: RIC

Duties and Responsibilities

- Meeting with aerospace engineers to discuss details and implications of test procedures
- Building and maintaining test facilities for aircraft systems
- Making and installing parts and systems to be tested in test equipment
- Operating and calibrating computer systems so that they comply with test and manufacturing requirements
- Making sure that test procedures are performed smoothly and safely
- Recording data from test parts and assemblies
- Installing instruments in aircraft and spacecraft
- Monitoring and ensuring quality in producing systems that go into the aircraft

A Day in the Life—Duties and Responsibilities

Aerospace engineering and operations technologists and technicians install, run, and maintain equipment used to develop, test, produce, and sustain aircraft and spacecraft. Their work is critical to ensuring the safety and precision of key parts of these vehicles and systems.

New aircraft designs undergo years of testing before they are put into service. As part of the job, technologists and technicians often calibrate test equipment, such as wind tunnels, and determine the causes of equipment malfunctions. They also may program and run computer simulations that test the new designs.

WORK ENVIRONMENT

Immediate Physical Environment

Aerospace engineering and operations technologists and technicians often work in offices. They may also work in manufacturing or industrial plants; in which case, they are frequently involved in assembly.

Most aerospace engineering and operations technologists and technicians work full time. Depending on the employer or project, they may work in shifts or be required to work overtime.

Human Environment

Aerospace engineering is not the solitary occupation one might expect, as they frequently work in engineering teams. Aerospace engineers may also be required to travel for work, be it to meet with stakeholders, or to go to testing grounds to test planes and spacecraft.

Technological Environment

Aerospace engineers and operations technicians use a multitude of tools in the course of their work and are versed in some of the most cutting-edge technologies available. Among other things, they are required to have knowledge of various operating system software such as Linux and/or UNIX; program testing software such as J Unit and Selenium; various programming languages, not to mention presentation software such as Microsoft PowerPoint. In addition, mechanical knowledge of machines and tools, including their designs, uses, repair, and maintenance is essential.

EDUCATION, TRAINING, AND ADVANCEMENT

High School/Secondary

High school students interested in becoming an aerospace engineering and operations technologist or technician should take classes in math, science, and, if available, drafting and information technology.

Suggested High School Subjects

- Algebra
- Geometry
- Trigonometry

- Pre-Calculus
- Physics

Related Career Pathways/Majors

Science, Technology, Engineering & Mathematics Career Cluster

- Engineering & Technology Pathway

Postsecondary

Aerospace engineering and operations technologists and technicians typically need an associate's degree in engineering technology or a related field. Some employers consider candidates who have a high school diploma or have completed a certificate program. Associate's degree and certificate programs are available at community colleges or vocational-technical schools.

Related College Majors

- Computer Engineering
- Electrical Engineering
- Electromechanical Engineering
- Mechanical Engineering
- Operations Research
- Robotics Engineering

Adult Job Seekers

Adults interested in becoming an aerospace engineer or operations technician should ideally have a high-school diploma at a minimum. Further experience, in the form of a certificate program, may be required in order to be familiar with the necessary processes and techniques.

Transferable Skills and Abilities

Communication Skills

- Following instructions from aerospace engineers, and clearly conveying problems to their supervisors

Detail-oriented

- Taking precise measurements needed by aerospace engineers, and keeping accurate records of these measurements

Interpersonal Skills

- Able to work well with others, and interacting with people from other divisions, businesses, and governments

Math Skills

- Using mathematics for measurement, analysis, design, and troubleshooting tasks

Mechanical Skills

- Assisting aerospace engineers by building what the engineers design, and guiding processes from design to production

Problem-solving Skills

- Troubleshooting design issues and evaluating system capabilities, formulating questions, and then finding the answers

Professional Certification and Licensure

Although not required for the job, SpaceTEC, the National Science Foundation's Center for Aerospace Technical Education, offers the Certified Aerospace Technician Core Certification. Recertification is required every three years.

Additional Requirements

In addition to a scientific educational background and strong mathematical skills, other highly useful skills for aerospace engineers and operations technicians include attention to detail and time management. Business acumen and communication skills are also highly valuable skills to have, as individuals in these professions will encounter instances where they need to liaise with clients.

EARNINGS AND ADVANCEMENT

The median annual wage for aerospace engineering and operations technologists and technicians was $68,570 in 2020. The lowest 10 percent earned less than $43,400, and the highest 10 percent earned more than $103,450.

Aerospace engineering and operations technologists may receive benefits such as paid vacations, holidays, and sick days; life and health insurance; and retirement benefits, usually paid by the employer.

EMPLOYMENT AND OUTLOOK

Employment of aerospace engineering and operations technologists and technicians is projected to grow 9 percent from 2020 to 2030, about as fast as the average for all occupations.

Most employment growth for these workers will be in the professional, scientific, and technical services industry.

Related Occupations

- Aircraft/Avionics Equipment Mechanic/Technician
- Drafter
- Electromechanical/Mechatronics Technologist/Technician
- Industrial Engineering Technologist/Technician
- Mechanical Engineering Technologist/Technician

MORE INFORMATION

Accreditation Board for Engineering and Technology, Inc. (ABET)
415 N. Charles Street
Baltimore, MD 21201
410.347.7700
comms@abet.org
www.abet.org

The American Society of Mechanical Engineers (ASME)
2 Park Avenue
New York, NY 10016-5990
800.843.2763
custcare@asme.org
www.asme.org

IEEE Aerospace and Electronic Systems Society
admin@ieee-aess.org
www.ieee-aess.org

International Association of Machinists and Aerospace Workers (IAMAW)
9000 Machinists Place
Upper Marlboro, MD 20772
301.967.4500
www.goiam.org

SpaceTEC
379 Cheney Highway
PMB 207
Titusville, FL 32780
321.567.5193
www.spacetec.us

Technology Student Association (TSA)
1904 Association Drive
Reston, VA 20191-1540
703.860.9000
general@tsaweb.org
www.tsaweb.org

Daniel Vest

Agricultural Engineer

Snapshot

Career Cluster(s): Agriculture, Food & Natural Resources; Science, Technology, Engineering & Mathematics

Interests: Problem Solving; Decision Making; Critical Thinking; Environmental Science; Mechanical Engineering; Farming

Earnings (Yearly Average): $84,410

Employment & Outlook: Slower Than Average Growth Expected

OVERVIEW

Sphere of Work

Agricultural engineers attempt to solve problems concerning power supplies, the efficiency of machinery, the use of structures and facilities, pollution and environmental issues, and the storage and processing of agricultural products.

Work Environment

Generally, agricultural engineers work in indoor office spaces. However, they may frequent a variety of worksites, both indoors and outdoors, such as laboratories or agricultural environments. Agricultural engineers with a focus in biology may spend more time doing lab work than field work.

Photo via iStock/Kinwun. [Used under license.]

Occupation Interest

Agricultural engineers have backgrounds in math, biology, physical science, and engineering. This career is ideal for someone with an interest in STEM as well as farming and the environment. A desire to work in a rural outdoor environment is beneficial. Depending on an agricultural engineer's additional interests, they may choose to specialize in a particular discipline related to the field. Those with a greater interest in the engineering aspect of this career may enjoy creating—and performing upkeep for—agricultural machinery. Conversely, someone more focused on plant biology may enjoy providing creative solutions to enhancing the quality and ensuring the health of crops and soil.

Profile

Working Conditions: Both Inside & Outside
Physical Strength: Light Work
Education Needs: Bachelor's Degree
Licensure/Certification: Optional
Opportunities for Experience: Internship; Apprenticeship
Interest Score: IRE

Duties and Responsibilities

- Using computer software to design equipment, systems, or structures
- Modifying environmental factors that affect animal or crop production, such as airflow in a barn or runoff patterns on a field
- Testing equipment to ensure its safety and reliability
- Overseeing construction and production operations
- Planning and working together with clients, contractors, consultants, and other engineers to ensure effective and desirable outcomes

A Day in the Life—Duties and Responsibilities

Agricultural engineers work in farming, including aquaculture (farming of seafood), forestry, and food processing. They work on a wide variety of projects. For example, some agricultural engineers work to develop climate control systems that increase the comfort and productivity of livestock whereas others work to increase the storage capacity and efficiency of refrigeration. Many agricultural engineers attempt to develop better solutions for animal waste disposal. Those with computer programming skills work to integrate artificial intelligence and geospatial systems into agriculture. For example, they work to improve efficiency in fertilizer application or to automate harvesting systems.

WORK ENVIRONMENT

Immediate Physical Environment

Agricultural engineers can work in both indoor office spaces and outdoor settings. When necessary, they will travel to agricultural settings to inspect the equipment and machinery and assure they are functioning according to both the manufacturers' specifications and federal and state regulations. Some agricultural engineers occasionally work in laboratories to test the quality of processing equipment or the effects that specific growing conditions have on plants. They may work onsite to supervise livestock facility upgrades or water resource management projects. Often, they will have to observe the results of their work where the crops are grown, or the livestock are held.

Agricultural engineers work with others in designing solutions to problems or applying technological advances. They work with people from a variety of backgrounds, such as business, agronomy, animal sciences, and public policy.

Agricultural engineers typically work full time. Schedules may vary because of weather conditions or other complications. When working on outdoor projects, agricultural engineers may work more hours to take advantage of good weather or fewer hours in case of bad weather. This profession has one of the highest rates of injuries and illnesses of all occupations.

In addition, agricultural engineers may need to be available outside of normal work hours to address unexpected problems that come up in manufacturing operations or rural construction projects.

Human Environment

Agricultural engineers often work directly with other engineers, scientists, and supervisors, and in some cases, clients. To work in this field, an agricultural engineer needs effective communication skills. They work with colleagues to collaboratively find solutions to imposing problems affecting crop production, livestock, water supply, waste disposal, and much more.

When working with clients, agricultural engineers need to be able to discuss the needs of the council, farmer, or developer in order to supply quality advice and decide the best approach to tackling their issues. In addition, field work may require agricultural engineers to explain complex problems in uncomplicated terms to their clients whose knowledge of biology or engineering may not be as extensive.

Technological Environment

Agricultural engineers must have an understanding of varying computer software, such as analytical and scientific programs, computer-aided design (CAD) platforms, graphic or photo imaging packages, geographic information systems, and typical office applications. In addition, they comprehend the mechanics that go into building or fixing large machinery and equipment. Agricultural engineers specializing in plant biology must know how to utilize laboratory technologies to analyze samples and data.

EDUCATION, TRAINING, AND ADVANCEMENT

High School/Secondary

Students interested in studying agricultural engineering will benefit from taking high school courses in math and science. Those who desire to study the biological aspects of agriculture should take advanced biology courses. Additional courses in English are helpful for sharpening communication skills. They should attempt to gain experience in the field through clubs and other extracurricular activities. If possible, working on a farm could give a student better insight into how livestock and crops are kept, as well as the common problems that affect them.

Suggested High School Subjects

- Algebra
- Biology
- Calculus
- Chemistry
- Civics
- Earth or Environmental Science
- Economics
- English
- Geography
- Geometry
- History
- Physics
- Statistics
- Trigonometry

Transferable Skills and Abilities

Analytical Skills
- Analyzing the needs of complex systems that involve workers, crops, animals, machinery and equipment, and the environment

Communication Skills
- Understanding the needs of clients, workers, and others working on a project
- Communicating ideas/recommendations about systems and about solutions to any problems being worked on

Math Skills
- Using calculus, trigonometry, and other advanced mathematical disciplines for analysis, design, and troubleshooting

Problem-solving Skills
- Solving problems found in agricultural production
- Designing safer equipment for food processing or reducing erosion
- Exploring creative ways of applying the principles of engineering

Related Career Pathways/Majors

Agriculture, Food & Natural Resources Career Cluster
- Power, Structural & Technical Systems Career Pathway

Science, Technology, Engineering & Mathematics Career Cluster
- Technology & Engineering Career Pathway

Postsecondary

University students interested in agricultural engineering should take courses in advanced calculus, physics, biology, and chemistry. They also may take courses in business, public policy, and economics.

Entry-level jobs in agricultural engineering require a bachelor's degree. Bachelor's degree programs in agricultural engineering or biological engineering typically include significant hands-on components in areas such as science, math, and engineering principles. Most colleges and universities encourage students to gain practical experience through projects such as participating in engineering competitions in which student teams design equipment and attempt to solve real problems.

Related College Majors

- Agriculture
- Biochemistry
- Biotechnology
- Engineering
- Food Science
- Soil Science

Adult Job Seekers

Adults seeking a career in agricultural engineering ideally have prior experience in a similar or related field. Those with experience in engineering and biology may have an easier time transitioning to a career in agricultural engineering. A bachelor's degree in a relevant subject is beneficial when trying to find work in this field. For those without a related degree or licensure that are interested in becoming an agricultural engineer, they must enroll in relevant continuing education courses or in a postsecondary accredited college or university before receiving their degree.

Professional Certification and Licensure

Licensure is not required for entry-level positions as an agricultural engineer. A Professional Engineering (PE) license, which allows for higher levels of leadership and independence, can be acquired later in one's career. Licensed engineers are called professional engineers (PEs). A PE can oversee the work of other engineers, sign off on projects, and provide services directly to the public. State licensure generally requires: a degree from an Accreditation Board for Engineering and Technology (ABET)-accredited engineering program; a passing score on the Fundamentals of Engineering (FE) exam; at least four years of relevant work experience; and a passing score on the Professional Engineering (PE) exam.

The initial FE exam can be taken after one earns a bachelor's degree. Engineers who pass this exam are commonly called engineers in training (EITs) or engineer interns (EIs). After meeting work experience requirements, EITs and EIs can take the second exam, called the Principles and Practice of Engineering (PE).

Fast Fact

With 914 million acres of farmland in the U.S. and world-class agricultural efficiency, the United States is the world's largest exporter of food.
Sources: www.realityworks.com and www.worldatlas.com

Photo via iStock/fotokostic. [Used under license.]

Each state issues its own licenses. Most states recognize licensure from other states, as long as the licensing state's requirements meet or exceed their own licensure requirements. Several states require engineers to take continuing education to keep their licenses. For licensing requirements, check with your state's licensing board.

Additional Requirements

Becoming an agricultural engineer requires an aptitude for creative problem- solving and an interest in finding ways to advance current technologies, systems, and structures to make them more productive. They must be comfortable working in various settings, including standard offices, laboratories, and farm environments.

EARNINGS AND ADVANCEMENT

Earnings of agricultural engineers depend on their education and experience, employer, and the types of responsibilities they take on. Median annual earnings of agricultural engineers were $84,410 in 2020. The lowest 10 percent earned less than $51,160, and the highest 10 percent earned more than $166,620.

Agricultural engineers may receive benefits such as paid vacations, holidays, and sick days; life and health insurance; and retirement benefits, usually paid by the employer.

Novice agricultural engineers usually work under the supervision of experienced engineers. As they gain knowledge and experience, beginning engineers move to more difficult projects and increase their independence in developing designs, solving problems, and making decisions.

With experience, agricultural engineers may advance to supervise a team of engineers and technicians. Some advance to become engineering managers. Agricultural engineers who become sales engineers use their engineering background to discuss a product's technical aspects with potential buyers and to help in product planning, installation, and use.

Agricultural engineers who have a master's degree or a PhD are more likely to be involved in research and development activities and may become postsecondary teachers.

EMPLOYMENT AND OUTLOOK

There were approximately 1,500 agricultural engineers employed nationally in 2020. Employment of agricultural engineers is expected to grow slower than the average for all occupations through the year 2030, at a rate of 5 percent.

Farming establishments will continue to require more machinery, equipment, and buildings to increase the efficiency of agricultural production systems and to reduce environmental damage, which should maintain demand for these workers.

Agricultural engineers are expected to continue working on projects such as alternative energies and biofuels; precision and automated farming technologies for irrigation, spraying, and harvesting; and growing food in space to support future exploration.

More efficient designs for traditional agricultural engineering projects such as irrigation, storage, and worker safety systems will also maintain demand for these workers. Growing populations and stronger global competition will result in the industry needing more efficient means of production, which will increase demand for agricultural engineers.

Related Occupations

- Agricultural/Food Science Technician
- Agricultural/Food Scientist
- Architectural/Engineering Manager
- Civil Engineer
- Conservation Scientist/Forester
- Environmental Engineer
- Farmer/Rancher/Agricultural Manager
- Hydrologist
- Industrial Engineer
- Mechanical Engineer

MORE INFORMATION

Accreditation Board for Engineering and Technology, Inc. (ABET)
415 North Charles St.
Baltimore, MD 21201
410.347.7700
comms@abet.org
www.abet.org

American Society for Engineering Education (ASEE)
1818 N Street N.W., Suite 600
Washington, DC 20036
202.331.3500
www.asee.org

American Society of Agricultural and Biological Engineers (ASABE)
2950 Niles Road
St. Joseph, MI 49085
269.429.0300
www.asabe.org

Future Farmers of America (FFA)
6060 FFA Drive
Indianapolis, IN 46278-1370
888.332.2668
www.ffa.org

National Council of Examiners for Engineering and Surveying (NCEES)
200 Verdae Boulevard
Greenville, SC 29607
800.250.3196
www.ncees.org

National Institute for Certification in Engineering Technologies (NICET)
1420 King Street
Alexandria, VA 22314
888.476.4238
socoffice@soc.org
www.nicet.org

National Institute of Food and Agriculture (NIFA)
United States Department of Agriculture
1400 Independence Avenue, SW, MS 2201
Washington, DC 20250-2201
nifa.usda.gov

National Society for Professional Engineers (NSPE)
1420 King Street
Alexandria, VA 22314
888.285.6773
www.nspe.org

Technology Student Association (TSA)
1904 Association Drive
Reston, VA 20191-1540
703.860.9000
general@tsaweb.org
www.tsaweb.org

U.S. Food and Drug Administration (FDA)
10903 New Hampshire Ave
Silver Spring, MD 20993-0002
888.463.6332
www.fda.gov

Emily Porter & Mark Boccard

Conversation With...
CAROLYN M. JONES, PE

Bay/Delta Team Agricultural Engineer
USDA Natural Resources Conservation Service (NRCS) California
In the field more than 20 years

What was your individual career path in terms of education/training, entry-level job, or other significant opportunity?

Growing up, I was involved in a trail-riding group, 4-H, FFA, and scouting with my parents that both had a strong "do it yourself" ethic—I'm pretty handy! As I entered junior high and high school, I excelled in math, the sciences, writing, and my vocational agricultural classes including my FFA Supervised Agricultural Experience Project (SAEP). Upon applying for college, my agricultural advisors encouraged me to consider agricultural and biological engineering (ABE). I went to the University of California at Davis and earned a B.S. in biosystems engineering (the ABE degree program name at the time) while also enjoying extra-curricular activities and classes in animal science and agriculture. A year prior to graduation, I learned about the USDA Natural Resources Conservation Service (NRCS) and earned an internship in a nearby field office working on stream, irrigation, pond, and other project types. Upon graduation, I joined the staff full-time and transitioned from trainee to field office engineer, and now to team engineer, where I assist staff and clients in seven offices with a wide variety of projects that address soil, water, air, plant, animal, and energy resource concerns.

What are the most important skills and/or qualities for someone in your profession?

Technical staff at USDA NRCS are specialized in agriculture, biology, and conservation. That also applies to the engineering work we do to support the nation's farmers, ranchers, forest landowners, and the public. We are a forward-facing agency, meaning we work directly with our clientele on their properties daily. Staff need to be professional and approachable, inquisitive about the client's natural resource concerns, and knowledgeable about the application of our Conservation Practice Standards on a variety of landscapes. And a gentle sense of humor is often beneficial in the workplace. As an engineer, I must be able to think critically, perform technical calculations and use programs such as computer-aided design (CAD), and work with our clients to identify viable alternatives and solve problems. Finally, it's important to acknowledge what you don't know—people will appreciate your honesty. But make sure you find out and follow up.

What do you wish you had known going into this profession?

I really love the work that I do, who I do it for, and why I do it! I also have found my skills in writing and documentation to be invaluable, to assure that design work and communications are clear, concise, and complete. Engineering is worth very little if it's not communicated well.

Are there many job opportunities in your profession? In what specific areas?

Since there will always be a need for agriculture to supply the population with food, fiber, and fuel, there will always be a role for engineers and other technical specialists in agriculture. As a part of a multi-disciplinary team, Agricultural and Biological Engineers can work across the breadth of agricultural and environmental systems to solve problems with good solutions. Agricultural and biological engineers solve problems using information and technology in the areas of plant, animal, and facility systems; machinery systems; natural resources and environmental systems; processing systems; ergonomics, safety, and health; applied sciences and engineering; to impact water, soil, air, food, fiber, and energy resources. The opportunities are endless with the appropriate attention to detail, problem solving skills, professionalism, and follow-through!

How do you see your profession changing in the next five years, how will technology shift, and what skills will be required?

Just like all other industries, agriculture continues to adopt new technologies! While we need to start by understanding why and how our clients do what they do to grow food, fiber, and fuel, there are many opportunities to incentivize the use of new and re-discovered technologies to reduce our collective environmental impact. A few examples include: expanded use of cover crops to improve soil health; better tracking of water deliveries and use with sensors and cloud-based data storage, and management for improved irrigation water application efficiency; and improved solid/liquid separation of dairy manure for better land application (as fertilizer) and biodigestion for electric energy production.

What do you enjoy most about your job? What do you enjoy least about your job?

The things I like most about my job are 1) helping our clients manage the balance of growing food, fiber, and fuel while reducing their impact on the environment; 2) working constructively with our clients to help them solve their identified problems; 3) teaching new staff and seeing their enthusiasm for the work we do (helping people help the land); and 4) the spectacular views and sights I get to see on project visits. I find my job the most difficult when conditions change for the client and the market for their crop, and a hard-worked design is not installed.

Can you suggest a valuable "try this" for students considering a career in your profession?

If you're interested in agricultural and biological engineering, I encourage you to study up in math, physics, and environmental science, but also seek out classes and extracurricular activities in vocational agriculture. Also consider reaching out to one of our colleagues through the American Society of Agricultural and Biological Engineers (ASABE) for an internship or opportunity to volunteer with their organization or company.

Conversation With...
GAYLE C. BAKER

Agricultural Engineer
Maurer-Stutz, Inc.
In the field, 12 years

What was your individual career path in terms of education/training, entry-level job, or other significant opportunity?
I received my Bachelor of Science in agricultural and biosystems engineering from Iowa State University. I worked in a research laboratory throughout my undergraduate studies, and had the opportunity to prepare several research papers and present at conferences and workshops. My first full-time job after graduation was with the United State Department of Agriculture Natural Resource Conservation Services (USDA-NRCS) in Illinois. This was a great opportunity and I focused on soil and water conservation design, and reviewed livestock housing and waste containment plans. I had the opportunity to work for a company that partnered with USDA-NRCS on livestock housing and management-facility planning, design, and construction.

What are the most important skills and/or qualities for someone in your profession?
Communication Skills—You would think that technical ability and skills would be first, but I have found that communication is what makes you an effective engineer and professional. Working as a team, communication with clients or external professionals, and written and verbal communication skills, convey your technical ability. Additionally, listening and comprehension skills are highly valued in this professional field.

Critical Thinking—Your ability to hear a problem and understand the details is the first part of solving the problem. The next part is your ability to think through different approaches and solutions to find the best, most efficient design or response.

Organization—Management of emails, files, and verbal communication is another important aspect of being an engineering professional. This makes you an effective team leader or project manager.

What do you wish you had known going into this profession?
When I went into this profession and my undergraduate studies, I only knew I wanted to work in agriculture and that I enjoyed critical thinking. In hindsight, some research into this a little further might have streamlined my education. When I started attending agricultural engi-

neering professional conferences and networking with other engineering professionals, I got a better understanding of what was possible in my career. My biggest advice is to get involved in organizations, meet people that can pass on information they have learned, and never be afraid to ask questions.

Are there many job opportunities in your profession? In what specific areas?

Yes, right now agricultural engineering is an expanding profession. Agriculture is actively pursuing unmanned and artificial intelligence solutions for production agriculture, and creating efficient farming techniques. Additionally, infrastructure is a priority nationwide, with opportunities to rehabilitate infrastructure such as dams and other land and water practices.

How do you see your profession changing in the next five years, how will technology shift, and what skills will be required?

Agricultural engineering is going to continue to move more and more into heavily data-based solutions for production agriculture. Creating smart solutions to streamline processes is a huge priority the agriculture industry.

What do you enjoy most about your job? What do you enjoy least about your job?

I enjoy that my job is very flexible. I have the opportunity to get out in the field to see projects in each phase of progress, from planning to construction, and in use by the client. I can also complete field work like surveying and soils investigations. A majority of my time is in the office, but it will range from calculations and design work, to communicating with clients, to drafting and preparing drawings.

Can you suggest a valuable "try this" for students considering a career in your profession?

Job-shadow and complete internships. This is an opportunity to see if the profession you are interested in is what you want to pursue. Another opportunity, if you are interested in developing design plans, is to help with construction. It will help you understand how to put together functional plans that communicate your design.

Conversation With...
REBECCA KALLAL

Design Engineer
Provisur Technologies
In the field, 7 years

What was your individual career path in terms of education/training, entry-level job, or other significant opportunity?

Early farming experience and experience with tools, machinery and project planning through 4-H. I also took Tech Ed and drafting classes in high school.

Undergrad Study:

- 4-year Bachelor of Science in Engineering at Purdue University. I studied in the Agricultural & Biological Engineering Department.

Summer internships during undergrad study:

- GSI—International leader in grain storage and in equipment for livestock farmers. I learned the practical use of computer-aided design (CAD) programs here, and I also expanded my understanding of how to create technical prints that can be used to manufacture goods.
- Caterpillar, Inc.—American Fortune 100 corporation that designs, develops, engineers, manufactures, markets, and sells machinery, engines, financial products, and insurance to customers via a worldwide dealer network. It is the world's largest construction-equipment manufacturer. I worked within the Six Sigma quality program, using Excel and other data tools to identify issues within the Caterpillar supply chain and help my managers make recommendations for solutions.

Study Abroad:

- 1 summer at the Universidad Nebrija in Madrid, Spain.

Student Clubs/Activities:

- EPICS (Engineering Projects in Community Service)
- International 1/4 Scale Tractor Design Competition
- All-American Marching Band

Graduate Study:

- I earned a Master's Degree in engineering management through an online program at the University of Wisconsin—Platteville. This allowed me to work full-time and finish my degree simultaneously. My company paid for my classes and course materials—this is a very common employer-offered benefit that is great to take advantage of.

1st Full-Time Engineering Position (7 years): Design engineer at Kuhn North America (International farm equipment manufacturer). I designed large-scale feed mixers for dairy and beef farmers.

2nd (Current) Engineering Position (less than 1 year): Mechanical engineer for Provisur Technologies (International food processing equipment manufacturer). I work on food processing equipment and processing lines—mostly for beef, chicken, alternative proteins, and pet food.

What are the most important skills and/or qualities for someone in your profession?

Being comfortable being hands-on with a project is a great skill to grow. Being able to display a willingness to work and "get your hands dirty" will earn you respect from those people you will be working with, or even managing in the future.

Also remember that communication is one of your greatest tools. This includes being able to give a clear presentation in front of small or large groups, and also being able to express your design intent through technical drawings or manuals.

What do you wish you had known going into this profession?

Being an engineer at some companies is like being a Swiss army knife—you are always prepared to do a little bit of everything. At other companies, you will be asked to focus all of your attention on specific area—to become a true specialist. Understanding which of those two types of engineer you want to be will help you weed out potential jobs during the interview process.

Are there many job opportunities in your profession? In what specific areas?

Yes. While not impervious to economic recession, I've noticed that careers in agriculture, including food processing, seem to rebound quickly. Especially if you are willing to broaden your search away from the brands that are the most recognizable, opportunities abound. Currently, the most abundant job openings are in the application of newer technology (computer programming, electrical engineering, etc.) to traditional processes.

How do you see your profession changing in the next five years, how will technology shift, and what skills will be required?

When we think of self-driving technology we are often thinking of the future of automotive engineering. However, tractors are adapting to use this technology today and will continue to advance in automation in the coming years. Farm equipment increasingly has more in common with your smart phone—such as touch screens, apps, and complex data management systems.

Likewise, in a time of unprecedented worker shortages, the labor-intensive jobs in food processing will become increasingly automated. Skills in programming, process automation, or high-tech manufacturing will help you land a job. The skills to learn and adapt to technology five, ten, twenty years from now will help you keep it!

What do you enjoy most about your job? What do you enjoy least about your job?
I enjoy the problem-solving aspect of engineering. Problem-solving extends beyond just designing product "X" to accomplish task "Y." It also encompasses designing products that can be manufactured in a timely, cost-efficient manner all while making sure those "efficiencies" don't impact your customers.

Can you suggest a valuable "try this" for students considering a career in your profession?
Become comfortable asking questions when you don't know how something works, and learning how to safely and intelligently approach new experiences. If you are someone who is interested in machinery, consider spending a day with a farmer (your high school's FFA or Tech Ed advisor may be able to help you there). If you are fascinated by learning how things are made, find out if local factories or food processing plants will offer tours. If an in-person tour is hard to find, the internet is full of videos showing start-to-finish processes. Take advantage of technical education classes at your school or local "Maker Spaces" that may be available to you.

Architectural/ Engineering Manager

Snapshot

Career Cluster(s): Science, Technology, Engineering & Mathematics

Interests: Architecture; Engineering; Science

Earnings (Yearly Average): $149,530

Employment & Outlook: Slower Than Average Growth Expected

OVERVIEW

Sphere of Work

Architectural and engineering managers plan, direct, and coordinate activities in architectural and engineering companies. Creating budgets, hiring staff, and consulting with other managers comprises a large part of their duties.

Work Environment

Architectural and engineering managers typically work in office environments, although others may perform laboratory work, or be required to attend industrial settings.

Occupation Interest

Architectural and engineering managers must possess vision, have strong leadership abilities, and know how to stay organized and focused. A creative mind and problem-solving abilities are essential, as not only do they create and spearhead new projects in their daily

Photo via iStock/Ngampol Thongsai. [Used under license.]

job, but they are the ones responsible for solving the problems inherent in each new undertaking. Managers should also be good communicators and judges of character as well, since they are also responsible for the hiring of team members, training those team members, and evaluating workers' performances.

Profile

Working Conditions: Both Inside and Outside
Physical Strength: Light Work
Education Needs: Bachelor's Degree
Licensure/Certification: N/A
Opportunities for Experience: Industry Work
Interest Score: ERI

A Day in the Life—Duties and Responsibilities

Architectural and engineering managers use their knowledge of architecture or engineering to oversee a variety of activities. They may direct and coordinate building activities at construction sites or activities related to production, operations, quality assurance, testing, or maintenance at manufacturing sites.

Architectural and engineering managers are responsible for developing the overall concept of a new product or for solving the technical problems that prevent the completion of a project. To accomplish this, they must determine technical goals and produce detailed plans.

Architectural and engineering managers spend a great deal of time coordinating the activities of their staff with the activities of other staff or organizations. They often confer with other managers, including those in finance, production, and marketing, as well as with contractors and equipment and materials suppliers.

Duties and Responsibilities

- Making detailed plans for the development of new products and designs
- Determining staff, training, and equipment needs
- Proposing budgets for projects and programs
- Hiring and supervising staff
- Leading research and development (R&D) projects to produce new products, processes, or designs
- Checking the technical accuracy of their staff's work
- Ensuring the soundness of methods their staff uses
- Coordinating work with other staff and managers

In addition, architectural and engineering managers must know how to prepare budgets, hire staff, and supervise employees. They propose budgets for projects and programs and determine staff, training, and equipment needs. These managers must also hire people and assign them specific parts of each project to carry out. Architectural and engineering managers supervise the work of their employees, set schedules, and create administrative procedures.

WORK ENVIRONMENT

Immediate Physical Environment

Most architectural and engineering managers work in offices, although some may also work in research laboratories and industrial production plants or at construction sites.

Most architectural and engineering managers work full time and some work more than 40 hours a week. These managers are often under considerable pressure to meet deadlines and budgets.

Human Environment

Architectural and engineering managers deal with others daily. They are responsible for hiring and supervising staff, evaluating staff performance. As well, conferring with other managers in different departments is a frequent necessity. Managers must also meet with contractors and suppliers on a regular basis.

Technological Environment

The technological requirements of architectural and engineering managers are many and can include but are not limited to: computer-aided design (CAD) software such as Revit and Rhino3D; analytical or scientific software; and database management system software such as Elastic Search and Apache Hive.

EDUCATION, TRAINING, AND ADVANCEMENT

High School/Secondary

Students interested in architectural and engineering manager careers should focus on science and math courses, as well as technical courses such as computer science. They should also consider joining their school's leadership or science clubs, if possible, for practical experience.

Suggested High School Subjects

- Algebra
- Calculus
- Geometry
- Physics
- Trigonometry

Related Career Pathways/Majors

Science, Technology, Engineering & Mathematics Career Cluster

- Engineering & Technology Pathway

Postsecondary

Architectural and engineering managers typically need at least a bachelor's degree in engineering or architecture.

Some also obtain business management skills by completing a master's degree in engineering management (MEM or MsEM) or technology management (MSTM) or a master's degree in business administration (MBA). Some workers earn their master's degree before advancing to management positions, and others earn it while they work as a manager. Typically, those who prefer to manage in technical areas pursue an MsEM or MSTM and those interested in more general management skills earn an MBA.

Engineering management programs usually include classes in accounting, engineering economics, financial management, industrial and human resources management, and quality control.

Technology management programs typically provide instruction in production and operations management, project management, computer applications, quality control, safety and health issues, statistics, and general management principles.

Related College Majors

- Architecture
- Business Administration
- Engineering
- Engineering Management
- Technology Management

Transferable Skills and Abilities

Analytical Skills
- Evaluating information carefully and solving complex problems

Communication Skills
- Overseeing staff and work together with other levels of management
- Communicating orders effectively and leading teams to meet goals

Detail-oriented
- Paying attention to detail
- Understanding complex systems, since a minor error can cause major problems

Math Skills
- Using calculus and other advanced mathematics to develop new products and processes

Organizational Skills
- Keeping track of many workers, schedules, and budgets simultaneously

Adult Job Seekers

Adults interested in going into architectural and engineering management will find that they need a bachelor's degree in either architecture, engineering, or business management. Managers only advance to their positions after years of employment as an architect or engineer. They usually have experience working on difficult or complex projects, developing designs, solving problems, and making decisions. Before moving up to a management position, they also typically gain experience leading engineering teams.

EARNINGS AND ADVANCEMENT

The median annual wage for architectural and engineering managers was $149,530 in 2020. The lowest 10 percent earned less than $95,310, and the highest 10 percent earned more than $208,000.

Architectural and engineering managers may receive benefits such as paid vacations, holidays, and sick days; life and health insurance; and retirement benefits, usually paid by the employer.

In addition, some architectural and engineering managers may receive more benefits—such as expense accounts and bonuses—than workers who are not managers.

EMPLOYMENT AND OUTLOOK

Employment of architectural and engineering managers is projected to grow 4 percent from 2020 to 2030, slower than the average for all occupations.

Employment growth will largely reflect the growth of the industries in which these managers are employed. Demand for civil engineering services is expected to continue as the nation's aging infrastructure requires expansion and repair. Mechanical and electrical engineering services should also be needed for projects such as wind turbine farms and other renewable energy construction and design.

Related Occupations

- Aerospace Engineer
- Architect
- Civil Engineer

Fast Fact

The Empire State Building, one of the most famous landlords in the world, actually generates more revenue from its observation decks than all of its office space combined.

Source: www.archdaily.com

Photo via iStock/tobiasjo. [Used under license.]

- Construction Manager
- Electrical and Electronics Engineer
- Industrial Engineer
- Industrial Production Manager
- Materials Engineer
- Mechanical Engineer
- Nuclear Engineer

MORE INFORMATION

Accreditation Board for Engineering and Technology, Inc. (ABET)
415 N. Charles Street
Baltimore, MD 21201
410.347.7700
comms@abet.org
www.abet.org

American Institute of Architects (AIA)
1735 New York Avenue, NW
Washington, DC 20006-5292
Fax: 202.626.7547
www.aia.org

American Society for Engineering Management (ASEM)
200 Sparkman Drive, Suite 2
Huntsville, AB 35805
256.503.8482
asem-hq@asem.org
www.asem.org

American Society of Civil Engineers (ACE)
1801 Alexander Bell Drive
Reston, VA 20191
800.548.2723
www.asce.org

The Association of Technology, Management, and Applied Engineering (ATMAE)
486 Cornell Road
Blairsville, PA 15717
919.635.8335
www.atmae.org

U.S. Green Building Council (USGBC)
2101 L Street, NW, Suite 500
Washington, DC 20037
800.795.1747
info@usgbc.org
www.usgbc.org

Daniel Vest

Bioengineer/ Biomedical Engineer

Snapshot

Career Cluster(s): Health Science; Science, Technology, Engineering & Mathematics

Interests: Inventing; Science; Medicine; Engineering; Consulting; Investigating

Earnings (Yearly Average): $92,620

Employment & Outlook: As Fast As Average Growth Expected

OVERVIEW

Sphere of Work

Bioengineers and biomedical engineers combine engineering principles with sciences to design and create equipment, devices, computer systems, and software.

Work Environment

Bioengineers and biomedical engineers typically work on teams with scientists, healthcare workers, investigators, or other engineers, depending upon the project. Engineers may be required to travel to locations where their creations are put into use, in order to gauge their effectiveness and note opportunities for improvement.

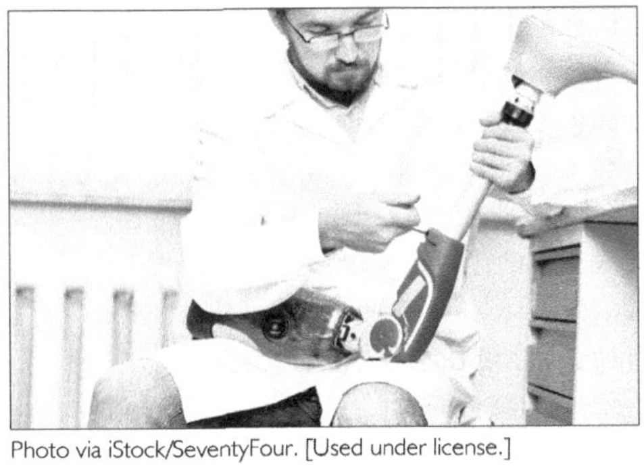

Photo via iStock/SeventyFour. [Used under license.]

Most bioengineers and biomedical engineers work full time, and some work more than 40 hours per week.

Profile

Working Conditions: Inside
Education Needs: Bachelor's Degree; Graduate Degree
Licensure/Certification: Usually Not Required
Opportunities for Experience: Internship; Co-op
Interest Score: IR

Occupation Interest

Bioengineers and biomedical engineers are creative, science-oriented individuals who have a desire to contribute to the scientific and medical fields through the creation of new tools and systems. They can identify gaps in the current market and have the technical expertise to design new equipment or software to fill those voids. They are eager to apply the latest technological advances towards new creations that will make scientific work faster, safer, and more efficient.

A Day in the Life—Duties and Responsibilities

Bioengineers and biomedical engineers often work in research and development (R&D) or quality assurance.

The work of bioengineers spans many fields. For example, although their expertise is in engineering and biology, they often design computer software to run complicated instruments, such as 3-D x-ray machines. Others use their knowledge of chemistry and biology to develop new drug therapies. Still others draw on math and statistics to understand signals transmitted by the brain or heart. Some are involved in sales.

Duties and Responsibilities

- Designing equipment and devices, such as artificial internal organs, replacements for body parts, and machines for diagnosing medical problems
- Installing, maintaining, or providing technical support for biomedical equipment
- Collaborating with manufacturing staff on the safety and effectiveness of biomedical equipment
- Working with scientists to research how engineering principles apply to biological systems
- Developing statistical models or simulations using statistical or modeling software
- Preparing procedures and writing technical reports and research papers
- Presenting research findings to a variety of audiences, including scientists, clinicians, managers, other engineers, and the public
- Designing or conducting follow-up experiments as needed

Biomedical engineers focus on advances in technology and medicine to develop new devices and equipment for improving human health. For example, they might design software to run medical equipment or computer simulations to evaluate new drug therapies. In addition, they design and build artificial body parts, such as hip and knee joints, or develop materials to make replacement parts. They also design rehabilitative exercise equipment.

OCCUPATION SPECIALTIES

Biochemical Engineer
Biochemical engineers focus on cell structures and microscopic systems to create products for bioremediation, biological waste treatment, and other uses.

Bioinstrumentation Engineer
Bioinstrumentation engineers use electronics, computer science, and measurement principles to develop tools for diagnosing and treating medical problems.

Biomaterials Engineer
Biomaterials engineers study naturally occurring or laboratory-designed substances for use in medical devices or implants.

Biomechanics Engineer
Biomechanics engineers study thermodynamics and other systems to solve biological or medical problems.

Clinical Engineer
Clinical engineers apply medical technology to improve healthcare.

Genetic Engineer
Genetic engineers alter the genetic makeup of organisms using recombinant deoxyribonucleic acid (rDNA) technology, such as in developing vitamin-fortified food crops to prevent disease in humans.

Rehabilitation Engineer
Rehabilitation engineers develop devices that aid people who are recovering from or adapting to physical or cognitive impairments.

Systems Physiologist
Systems physiologists use engineering tools to understand how biological systems function and respond to changes in their environment.

WORK ENVIRONMENT

Immediate Physical Environment

Bioengineers and biomedical engineers work on teams with scientists, healthcare workers, investigators, or other engineers. Where and how they work depends upon the project. For example, a biomedical engineer who has developed a new device might spend hours in a hospital to ensure that the device works as planned. If the device needs adjusting, the engineer might need to suggest alterations in the manufacturing process.

Human Environment

As engineering work is usually a team effort, bioengineers and biomedical engineers must be prepared to collaborate with colleagues, sometimes from varying disciplines. They may also interact with clients and patients during the design and implementation stages, including on-site testing of equipment. Consulting engineers must be prepared to work with investigators in the field, sometimes under stressful conditions.

Technological Environment

Engineers must be able to incorporate the latest technological advances into their work, to further their own creations and better the scientific and medical communities. Nanotechnology, smartphone technology, three-dimensional printing, and artificial intelligence (AI) are all examples of constantly evolving technologies that can play a role in bioengineering. These engineers must continue to expand their knowledge as technology progresses.

EDUCATION, TRAINING, AND ADVANCEMENT

High School/Secondary

In high school, students interested in becoming bioengineers or biomedical engineers should take classes in sciences such as chemistry, physics, and biology. They should also study math, including algebra, geometry, trigonometry, and calculus. If available, classes in drafting, mechanical drawing, and computer programming are also useful.

Suggested High School Subjects

- Algebra
- Biology
- Chemistry
- Civics
- Computer Science
- Drafting
- Earth or Environmental Science
- Economics

- English
- Entrepreneurship
- Geometry
- History
- Physics
- Pre-Calculus
- Psychology
- Statistics
- Trigonometry

Related Career Pathways/Majors

Health Science Career Cluster

- Biotechnology Research & Development Pathway

Science, Technology, Engineering & Mathematics Career Cluster

- Engineering & Technology Pathway
- Science & Mathematics Pathway

Transferable Skills and Abilities

Analytical Skills
- Assessing the needs of patients and customers prior to designing products

Communication Skills
- Working with patients, customers, and often working on teams
- Writing reports and research papers

Creativity
- Conceiving innovations in healthcare equipment and devices

Math Skills
- Using calculus and other advanced math and statistics for analysis, design, and troubleshooting in their work

Mechanical Skills
- Operating complex machinery and occasionally performing routine maintenance

Problem-solving Skills
- Dealing with intricate biological systems
- Working independently and with others to incorporate ideas into the complex problem-solving process

Postsecondary

At the bachelor's degree level, prospective bioengineers should enter bioengineering or traditional engineering programs, such as mechanical and electrical. Students who pursue traditional engineering degrees may benefit from taking biological science courses.

Bachelor's degree programs in bioengineering and biomedical engineering focus on engineering and biological sciences. These programs typically include laboratory- and classroom-based courses in biological sciences and subjects such as fluid and solid mechanics, circuit design, and biomaterials.

These programs also include substantial training in engineering design. As part of their study, students may have an opportunity to participate in co-ops or internships with hospitals and medical device and pharmaceutical manufacturing companies. Bioengineering and biomedical engineering programs are accredited by the Accreditation Board for Engineering and Technology, Inc. (ABET).

Related College Majors

- Biochemical Engineering
- Biological Engineering
- Chemical Engineering

Fast Fact

Biomedical engineering has one of the highest percentages of female students in any engineering field. Many programs also offer joint bachelor's and master's programs to encourage women to pursue doctorate degrees. Source: healthresearchfunding.org

Photo via iStock/katleho Seisa. [Used under license.]

- Computer Engineering
- Electrical Engineering
- Engineering Science
- Materials Engineering
- Mechanical Engineering
- Polymer & Plastics Engineering
- Robotics Engineering

Adult Job Seekers

Some bioengineers attend medical or dental school to specialize in techniques such as using electric impulses in new ways to get muscles moving again. Others earn law degrees and work as patent attorneys. Still others pursue a Master of Business Administration (MBA) degree and move into managerial positions.

EARNINGS AND ADVANCEMENT

Earnings depend upon the type and scale of projects undertaken. Median annual earnings of bioengineers or biomedical engineers were $92,620 in 2020. The lowest 10 percent earned less than $56,590, and the highest 10 percent earned more than $149,440.

Bioengineers or biomedical engineers may receive paid vacations, holidays, and sick days; life and health insurance; and retirement benefits. These are usually paid by the employer. Bioengineers or biomedical engineers may also be able to expense travel to on-site locations for product testing.

Bioengineers and biomedical engineers may increase their responsibilities as they gain experience or advanced degrees. To lead a research team, a bioengineer or biomedical engineer typically needs a graduate degree. Those who are interested in basic research may become medical scientists.

EMPLOYMENT AND OUTLOOK

Bioengineers and biomedical engineers held 19,300 jobs in 2020. Employment is expected to grow as fast as average for all occupations through the year 2029, at a rate of 6 percent, because of increasing technologies and their applications to medical equipment and devices. Smartphone technology and three-dimensional printing are examples of technology being applied to biomedical advances.

As the baby-boom generation lives longer and stays active, the demand for bioengineers and biomedical devices and procedures, such as hip and knee replacements, is expected to increase. In addition, as the public awareness of medical advances continues, increasing numbers of people will seek biomedical solutions to their health problems from their physicians.

Bioengineers and biomedical engineers work with scientists, other medical researchers, and manufacturers to address a range of injuries and physical disabilities. The ability of these engineers to collaborate on activities with workers from other fields is enlarging the range of applications for biomedical engineering products and services.

Related Occupations

- Agricultural Engineer
- Architectural/Engineering Manager
- Biochemist/Biophysicist
- Chemical Engineer
- Electrical/Electronics Engineer
- Materials Engineer
- Mechanical Engineer
- Medical Scientist
- Physician/Surgeon
- Sales Engineer

MORE INFORMATION

Accreditation Board for Engineering and Technology, Inc. (ABET)
415 North Charles Street
Baltimore, MD 21201
410.347.7700
www.abet.org

American Association of Engineering Societies (AAES)
1801 Alexander Bell Drive
Reston, VA 20191
202.296.2237
orders@aaes.org
www.aaes.org

American Institute for Medical and Biological Engineering (AIMBE)
1400 I Street NW, Suite 235
Washington, DC 20005
202.496.9660
info@aimbe.org
aimbe.org

American Society for Engineering Education (ASEE)
1818 N Street NW, Suite 600
Washington, DC 20036
202.331.3500
www.asee.org

American Society of Agricultural and Biological Engineers (ASABE)
2950 Niles Road
St. Joseph, MI 49085
269.429.0300
www.asabe.org

American Society of Mechanical Engineers (ASME)
Two Park Avenue
New York, NY 10016-5990
800.843.2763
CustomerCare@asme.org
www.asme.org

Biomedical Engineering Society (BMES)
8201 Corporate Drive, Suite 1125
Landover, MD 20785
membership@bmes.org
301.459.1999
www.bmes.org

Engineering Education Service Center (EESC)
1411 Old Hardman Btms Road
Clarkesville, GA 30523
706.499.5011
www.engineeringedu.com

IEEE Engineering in Medicine and Biology Society (EMBS)
445 Hoes Lane
Piscataway, NJ 08854
732.981.3433
www.embs.org

Institute of Biological Engineering (IBE)
446 East High Street, Suite 10
Lexington, KY 40507
859.977.7450
info@ibe.org
www.ibe.org

Society of Biological Engineers (SBE)
c/o American Institute of Chemical Engineers (AIChE)
120 Wall Street, FL 23
New York, NY 10005-4020
800.242.4363
www.aiche.org/sbe

Technology Student Association (TSA)
1904 Association Drive
Reston, VA 20191-1540
703.860.9000
general@tsaweb.org
tsaweb.org

Stuart Paterson

Conversation With...
RICHARD COTE

President and Founder
Avidien Technologies, Inc., Hudson, MA
Product design and development, development management, business management, 26 years

What was your individual career path in terms of education/training, entry-level job, or other significant opportunity?
My father worked as a technical instructor at Massachusetts Institute of Technology, and we had an extensive shop in the basement; so, I was exposed to machine shops and how to put things together. After high school, I spent about 18 months doing different jobs like working on a commercial fishing boat. I also got to the point I could not only take apart a car engine but understand the technical side of it. That led me to want to study mechanical engineering.

While studying mechanical engineering at Northeastern University in the early 1990s, I took advantage of their cooperative education program. In my final co-op assignment, I worked for a company called Costar in Cambridge, Massachusetts, that made life science laboratory equipment and disposable labware.

The exposure in that initial job was the start of my career developing new products used in life science labs around the world. The guy I worked for during my last co-op session offered me a job, and I worked for him or with him for 20 years. He and I moved from one company to another, founded a company, and he is still a good friend.

I was at Costar for about 6 years, then moved to Matrix Technologies. Due to a merger, our parent company was bought. Five of us decided to start a new company, Viaflow Integra. That company is now very successful, but it was both challenging and exciting in the beginning. That experience led me to start my own company, Avidien Technologies Inc. I spent a couple of years doing product development in my spare time while working for another company before launching Avidien full-time.

If you work in chemistry or life sciences, you have to be able to transfer liquids precisely. We make the smallest of a type of automated pipette that is really easy and intuitive to use. Automation reduces mistakes. There are competitors out there, but we developed many concepts that allow these instruments to be small.

We spent 5 years creating the company and designing our product, the first of which was released in 2017. Our clients include pharmaceutical and diagnostic companies, as well as university labs.

Many entrepreneurs have an idea and then go find their funding first. Avidien was bootstrapped, meaning that the company was self-funded through savings, family, and friends. The main reason I was able to afford to bootstrap the company is because I could do a lot of the engineering work myself.

There are many paths that can lead to entrepreneurial pursuits, but for me the cooperative education model followed by direct exposure to the startup environment was the course that led me to where I am today. In addition, most new companies are built around new products, so being an engineer involved in the new product development process before starting Avidien gave me a good foundation on which to build a company.

What are the most important skills and/or qualities for someone in your profession?
As an entrepreneur, you have to be willing to take risks and push yourself to keep trying when failure is always a possibility. Above all, the main quality an entrepreneur has to possess is a deep belief in themselves and their ideas.

What do you wish you had known going into this profession?
Starting a business requires the founders and early employees to wear many hats—marketing, sales, human resources, accounting, etc. In most jobs people tend to stay in their professional lanes and not pay much attention to the other business functions around them. Through my career I wish that I had paid more attention and asked more questions that in retrospect would have helped me while starting Avidien.

Also, given hindsight, I would tell somebody thinking about entrepreneurship to try it early on. Not so early you don't have enough work experience, but early enough so you're not deep into a mortgage and family. The risk is lower. Also, accept you may fail once or twice but you learn from the failure.

Are there many job opportunities in your profession? In what specific areas?
Anyone with the right drive and determination can choose to be an entrepreneur; so, it is not really about job opportunity. However, the key to being a successful entrepreneur is choosing a specific area that is a fit for your skills and your passions, and that is also a marketable endeavor, whether a service, product, or other enterprise.

How do you see your profession changing in the next 5 years? How will technology impact that change, and what skills will be required?
In the next five years it will continue to become easier to be a successful entrepreneur. For example, finding your customers is a key aspect to starting a business. As technology advances, reaching potential customers worldwide will continue to evolve into something you can do simply through your computer and internet connection. Success will go to those who develop skills marketing products through the constantly evolving social media platforms and are able to take advantage of that access.

What do you enjoy most about your job? What do you enjoy least about your job?
The aspect I enjoy most about being an entrepreneur in a company that develops products for laboratory use is hearing positive feedback about our company's products and service from our customers. Fortunately, this happens frequently. Hearing about a problem a customer is having with one of our products is something I enjoy least. On the positive side, it is also how we learn to improve our products.

Can you suggest a valuable "try this" for students considering a career in your profession?
If a student is interested in pursuing a life as an entrepreneur, I would recommend that they work or spend some time in a young startup company to get a flavor for the environment and its challenges. There are also many online and print periodicals dedicated to the aspects of becoming a successful entrepreneur that will paint a bigger picture of what it means to start a company. I also think the show Shark Tank is great for understanding the fundraising component and how investors think about a new venture.

This interview was originally published in 2021.

Conversation With...
BEN NOE

Research & Development Project Manager
Medtronic, Boulder, CO
Biomedical Engineer, 21/2 years

What was your individual career path in terms of education/training, entry-level job, or other significant opportunity?

I have a Bachelor of Science in mechanical engineering from the University of Colorado at Boulder and a certification in project management through the Project Management Institute. Even though my formal training is in mechanical engineering, biomedical engineering was my focus in college. My college didn't have a formal degree program in it. I had considered becoming a doctor or a physical therapist, but as a freshman or sophomore in college, I came to understand my strengths were in engineering and business.

My entry-level job is as an Associate Project Manager with Medtronic's Minimally Invasive Therapies Group. I had been a project management intern at Covidien and transitioned into full-time. I was lucky in that there was a need for my specific position when I graduated and I was hired right out of college. Medtronic acquired Covidien, so now I work for Medtronic. I'm heading my own fairly large project right now and I'm pretty excited about it—an electrosurgical generator that controls electrical cutting and coagulation instruments in surgery.

Another significant opportunity was my involvement with the Biomedical Engineering Society. I started a university chapter of the society when I was in college. That helped with me get my internship because I made contact with a lot of local companies. Also, leading that team helped in my interview for the internship. Now I've moved up to Industry Committee Chair and am on the Board of Directors for the society.

What are the most important skills and/or qualities for someone in your profession?

I think the most important skills are the ability to understand processes and regulations that are the framework in which we're working. The medical device industry, and really biomedicine in general, is heavily regulated. Also, the ability to work together in a team to come up with innovative solutions is a very important skill.

What do you wish you had known going into this profession?

I wish I had a better background in quality processes/quality control, regulations and restrictions, and then manufacturing. Some universities do offer these types of classes, for example,

"Design Control in Medical Devices" or "The Regulatory Environment in the Biomedical Field." They're very specialized classes.

Are there many job opportunities in your profession? In what specific areas?

Absolutely. It's a quickly growing field. The majority of biomedical engineering positions are in academia and industry. There are some government jobs, for instance with the Food and Drug Administration (FDA) or National Institutes of Health (NIH).

How do you see your profession changing in the next five years? What role will technology play in those changes, and what skills will be required?

I see three key things in terms of the profession changing. One is the consolidation of companies into several large companies. A lot of the smaller and mid-size companies, and even some larger companies, are being gobbled up by a few major companies. The second thing is integration with high tech. There's been a huge push to work with Apple or Google on things like personalized health information on an iPhone. And the third one would be a move toward more automation and robotic procedures, even in an operating room, where everything might be controlled from an iPad. The benefit can be a lot more precision and the possibility of eliminating human error.

What do you enjoy most about your job? What do you enjoy least about your job?

What I enjoy most, honestly, is seeing the product in action and seeing the impact our products have on patients' lives. Working in electrosurgery, where you can cut and cauterize tissue with better efficiency than a suture and scalpel, seeing the results for patients in terms of faster recovery time and less pain has been a rewarding experience.

What I enjoy least is probably the conflict management part of the job. A lot of the conflict is around interpretation of and different approaches to applying standards and regulations. Many of the standards are somewhat vague—actually, intentionally somewhat vague—and there can be disagreement over how they apply to our products.

Can you suggest a valuable "try this" for students considering a career in your profession?

My advice would be to try to get an internship. You could also start a club or seek out a mentor program. Reach out to someone in the field, just to be able to get on a job for one day. Some companies offer what they call an externship, where you can shadow an engineer for a day.

This interview was originally conducted in 2015.

Chemical Engineer

Snapshot

Career Cluster(s): Science, Technology, Engineering & Mathematics

Interests: Chemistry; Mathematics; Problem—solving; Critical Thinking; Analyzing Data

Earnings (Yearly Average): $108,540

Employment & Outlook: As Fast As Average Growth Expected

OVERVIEW

Sphere of Work

Chemical engineers apply the principles of chemistry, biology, physics, and math to solve problems that involve the production or use of chemicals, fuel, drugs, food, and many other products. They design processes and equipment for large-scale manufacturing, plan and test production methods and byproducts treatment, and direct facility operations.

Work Environment

Chemical engineers generally work in offices or laboratory settings, although sometimes they must work in an industrial setting to oversee production.

Occupation Interest

Chemical engineers usually perform well in science and mathematics. However, this field is fit for both creative and logical thinkers. This profession creatively uses the concepts of chemistry, biology, physics, and math to come up with a variety of solutions to real-world issues and obstacles. Chemical

Photo via iStock/industryview. [Used under license.]

engineers use their problem-solving skills and analytical minds to discover and fix problems and improve existing products and technologies. Someone who wants to make positive changes for people or the environment, who is also not opposed to time spent conducting research and analyzing data, may find fulfillment in this profession.

Profile

Working Conditions: Inside Work
Physical Strength: Light Work
Education Needs: Bachelor's Degree
Licensure/Certification: Optional
Opportunities for Experience: Internship, Apprenticeship
Interest Score: IR

A Day in the Life—Duties and Responsibilities

Chemical engineers develop and design chemical manufacturing processes. Some chemical engineers, known as process engineers, specialize in a particular process, such as oxidation (a reaction of oxygen with chemicals to make other chemicals) or polymerization (making plastics and resins).

Others specialize in a particular field, such as nanomaterials (extremely small substances) or biological engineering. Still others specialize in developing specific products.

In addition, chemical engineers work in the production of energy, electronics, food, clothing, and paper. They must understand how the manufacturing process affects the environment and the safety of workers and consumers.

Duties and Responsibilities

- Conducting research to develop new and improved manufacturing processes
- Establishing safety procedures for those working with dangerous chemicals
- Developing processes for separating components of liquids and gases, or for generating electrical currents, by using controlled chemical processes
- Designing and planning the layout of equipment
- Conducting tests and monitoring the performance of processes throughout production
- Troubleshooting problems with manufacturing processes
- Evaluating equipment and processes to ensure compliance with safety and environmental regulations
- Estimating production costs for management

Chemical engineers also conduct research in the life sciences, biotechnology, and business services.

WORK ENVIRONMENT

Immediate Physical Environment

Chemical engineers work mostly in offices or laboratories. They may spend time at industrial plants, refineries, and other locations, where they monitor or direct operations or solve on—site problems. Chemical engineers must be able to work with those who design other systems and with the technicians and mechanics that put the designs into practice. Some engineers travel extensively to plants or worksites, both domestically and abroad.

Chemical engineers can be exposed to health or safety hazards when handling certain chemicals and plant equipment, but such exposure can be avoided if proper procedures are followed.

Nearly all chemical engineers work full time. Occasionally, they may have to work additional hours to meet production targets and design standards or to troubleshoot problems with manufacturing processes. Some chemical engineers work more than 40 hours per week.

Human Environment

While some chemical engineers may conduct isolated work within their office or laboratory setting, many communicate with a wide range of individuals on a daily basis, from other engineers and lab techs to government officials, construction personnel, and machine operators. Developing healthy workplace relationships with coworkers is important to ensure practical communication between a team of engineers working on a project.

Technological Environment

Chemical engineers use various computer software to complete their work, including analytical and scientific software, computer—aided design (CAD) software, and office suite software. In addition to any equipment that they may design themselves, chemical engineers may use freeze dryers, heat exchangers, laboratory mixers, spectrometers, and microscopes of various sizes. They may operate plant equipment or use laboratory gear to handle potentially hazardous chemicals, which must be safely maneuvered to avoid harmful accidents or exposure.

EDUCATION, TRAINING, AND ADVANCEMENT

High School/Secondary

High school students interested in studying chemical engineering should take science classes, such as chemistry, physics, and biology. They also should take math courses, including algebra, trigonometry, and calculus. English courses are very valuable for sharpening communication and interpersonal skills. Because chemical engineers use CAD software, computer programming courses may give student a jumpstart on understanding these complex computer systems. They can gain beginner experience in the field through engineering, science, or math clubs and other extracurricular activities.

Suggested High School Subjects

- Algebra
- Biology
- Calculus
- Chemistry
- Civics
- Computer Programming
- Economics
- English
- Entrepreneurship
- Geography
- Geometry
- History
- Math Analysis
- Physics
- Statistics
- Trigonometry

Related Career Pathways/Majors

Science, Technology, Engineering & Mathematics Career Cluster

- Engineering & Technology Career Pathway

Transferable Skills and Abilities

Analytical Skills

- Troubleshooting designs that do not work as planned
- Asking the right questions and finding answers that work

Creativity

- Exploring new ways of applying engineering principles
- Working to invent new materials, advanced manufacturing techniques, and new applications in chemical and biomedical engineering

Ingenuity

- Learning the broad concepts of chemical engineering and applying those concepts to specific production problems

Interpersonal Skills

- Developing good working relationships with other workers involved in production processes

Math Skills

- Using the principles of advanced math topics such as calculus for analysis, design, and troubleshooting in their work

Problem-solving Skills

- Anticipating and identifying problems, including such issues as workers' safety and problems related to manufacturing and environmental protection

Postsecondary

Chemical engineers must have a bachelor's degree in chemical engineering or a related field. Programs in chemical engineering usually take four years to complete and include classroom, laboratory, and field studies.

At some universities, students can opt to enroll in five—year engineering programs that lead to both a bachelor's degree and a master's degree. A graduate degree, which may include a degree up to the PhD level, allows an engineer to work in research and development (R&D) or as a postsecondary teacher.

Some colleges and universities offer internships and/or cooperative programs in partnership with industry. In these programs, students gain practical experience while completing their education.

Related College Majors

- Biochemistry
- Biotechnology
- Chemical Engineering
- Chemistry
- Clinical Laboratory Science
- Engineering
- Forensic Chemistry
- Materials Engineering
- Pharmaceutical Science

Adult Job Seekers

Ideally, adults seeking a career in chemical engineering have experience in a similar or related field. A bachelor's degree in a relevant subject is necessary when trying to find work in this field. Adult job seekers without a related degree or applicable experience that are interested in becoming a chemical engineer need to enroll in relevant continuing education courses or in a postsecondary accredited college or university. Without receiving a bachelor's degree in chemical engineering or a similar subject, there are no opportunities for an adult to work in this field.

Professional Certification and Licensure

Licensure for chemical engineers is not as common as it is for other engineering occupations, nor is it required for entry-level positions. A Professional Engineering (PE) license, which allows for higher levels of leadership and independence, can be acquired later in one's career. Licensed engineers are called professional engineers (PEs). A PE can oversee the work of other engineers, sign off on projects, and provide services directly to the public. State licensure generally requires a degree from an Accreditation Board for Engineering and Technology, Inc. (ABET)—accredited engineering program; a passing score on the Fundamentals of Engineering (FE) exam; relevant work experience, typically at least four years; and a passing score on the Professional Engineering (PE) exam.

The initial FE exam can be taken after one earns a bachelor's degree. Engineers who pass this exam are commonly called engineers in training (EITs) or engineer interns (EIs). After meeting work experience requirements, EITs and EIs can take the second exam, called the Principles and Practice of Engineering (PE).

Fast Fact

Chemical engineers are also involved in space exploration, including the design of fuel cells and the manufacture of propellants. In the future, they may also be able to solve crucial problems such as the adequate supply of fresh water around the world through desalination of seawater, and advanced pollution prevention. Source: www.britannica.com

Photo via iStock/ArtTim. [Used under license.]

Each state issues its own licenses. Most states recognize licensure from other states, as long as the licensing state's requirements meet or exceed their own licensure requirements. Several states require engineers to take continuing education to keep their licenses.

Additional Requirements

In addition to having an advanced understanding of science and math, chemical engineers must have a willingness to provide creative solutions to real-world problems, find ways to make advancements on existing products, or create new products altogether.

EARNINGS AND ADVANCEMENT

Earnings of chemical engineers depend on their education and experience, employer, licensure, and the types of responsibilities they take on. Median annual earnings of chemical engineers were $108,540 in 2020. The lowest 10 percent earned less than $68,430 and the highest 10 percent earned more than $168,960.

Chemical engineers may receive benefits such as paid vacations, holidays, and sick days; life and health insurance; and retirement benefits, usually paid by the employer.

Entry-level engineers usually work under the supervision of experienced engineers. In large companies, new engineers also may receive formal training in classrooms or seminars. As junior engineers gain knowledge and experience, they move to more difficult projects with greater independence to develop designs, solve problems, and make decisions.

Eventually, chemical engineers may advance to supervise a team of engineers and technicians. Some may become architectural and engineering managers. Preparing for management positions usually requires working under the guidance of a more experienced chemical engineer.

EMPLOYMENT AND OUTLOOK

There were approximately 26,300 chemical engineers employed nationally in 2020. Employment of chemical engineers is expected to grow as fast as average for all occupations through the year 2030, at a rate of 9 percent.

Demand for chemical engineers' services depends largely on demand for the products of various manufacturing industries. Many chemical engineers work in manufacturing firms that provide products to other firms. For example, environmental and sustainability concerns have led chemistry and manufacturing firms to research alternative fertilizers, resulting in a need for chemical engineers.

In addition, chemical engineering will continue to migrate into other fields, such as nanotechnology, alternative energies, and biotechnology, and thereby help to sustain demand for engineering services in many manufacturing industries.

Related Occupations

- Architectural/Engineering Manager
- Bioengineer/Biomedical Engineer
- Chemical Technician
- Chemist/Materials Scientist
- Nuclear Engineer
- Occupational Health & Safety Specialist/Technician

MORE INFORMATION

Accreditation Board for Engineering and Technology, Inc. (ABET)
415 N. Charles Street
Baltimore, MD 21201
410.347.7700
comms@abet.org
www.abet.org

American Institute of Chemical Engineers (AIChE)
120 Wall Street, Floor 23
New York, NY 10005-4020
800.242.4363
www.aiche.org

American Institute of Chemical Engineers (AIChE) Career Center
120 Wall Street, Floor 23
New York, NY10005—4020
646.495.1321
internships.aiche.org

American Society for Engineering Education (ASEE)
1818 N Street, NW, Suite 600
Washington, DC 20036
202.331.3500
www.asee.org

National Council of Examiners for Engineering and Surveying (NCEES)
200 Verdae Boulevard
Greenville, SC 29607
800.250.3196
www.ncees.org

National Society of Professional Engineers (NSPE)
1420 King Street
Alexandria, VA 22314
888.285.6773
www.nspe.org

Technology Student Association (TSA)
1904 Association Drive
Reston, VA 20191—1540
703.860.9000
general@tsaweb.org
www.tsaweb.org

Conversation With...
ANDREW ZYDNEY

Distinguished Professor, Chemical Engineering
Pennsylvania State University
Chemical Engineer, 35 years

What was your individual career path in terms of education/training, entry-level job, or other significant opportunity?
I received my B.S. in Chemical Engineering from Yale University, then went to M.I.T. for my Ph.D. in Chemical Engineering. I had always been interested in-and good at-science and math, and thought engineering might be a good fit. I didn't know much about the field, but took an "Introduction to Chemical Engineering" course as a sophomore, and did so well that the instructor offered me a position as a research assistant in his lab. I loved doing research, and I became hooked on the discipline as a career path. I've never regretted that. I enjoy applying the principles of chemistry to solve problems, as well as the combination of mathematics with chemistry.

In graduate school, I was involved in a research project looking at the use of membrane filtration to remove toxic plasma proteins from blood, which was being considered as a treatment for diseases ranging from cancer to rheumatoid arthritis to lupus. I really enjoyed learning the fundamentals of membrane separations and was excited about the impact the work could have on human health. I have continued to do research in both membranes and health-related technology. I'm currently working on projects ranging from the purification of polysaccharide-based vaccines, to the use of novel membranes for the purification of plasmid DNA for gene therapy, to the analysis of virus removal filtration processes used to ensure the safety of a wide range of therapeutics.

What are the most important skills and/or qualities for someone in your profession?
My roles as chemical engineer and professor each have a somewhat different, although overlapping, skill set. Chemical engineers need a solid foundation in chemistry, physics, and the life sciences, as well as a strong background in math. They need to be good problem solvers who see the big picture.

As a professor, I have significant responsibilities in research and teaching. Both require strong communication skills, although the nature of the communications is very different. Research requires creativity and insight, including the ability to draw connections between ideas. As a researcher, I spend a tremendous amount of time writing manuscripts, funding proposals, and progress reports.

Teaching requires an ability to recognize that everyone has his or her own learning style. My responsibility is to try to connect with each student and effectively communicate complex material in a way that expands each student's understanding.

What do you wish you had known going into this profession?
I initially had no idea of the field's breadth. Chemical engineers provide a unique perspective to the industries in which they're involved, with a focus on the chemical and/or biological reactions involved in making and purifying products and the molecular characteristics of novel materials and products.

I also did not appreciate the importance of strong communications skills; I naively thought all I needed was a strong technical base.

Are there many job opportunities in your profession? In what specific areas?
Chemical engineers find jobs in the pharmaceutical and biotechnology industries, medicine, electronics, plastics, the food industry, consumer products, consulting, the environment, biofuels and renewable energy, nanotechnology, process safety, and the chemical and petroleum industries. Recent natural gas finds in the U.S. have created exciting new opportunities in energy.

How do you see your profession changing in the next five years, what role will technology play in those changes, and what skills will be required?
Developing more sustainable sources of energy and converting life sciences discoveries into cost-effective approaches for improving human health are key challenges. Solutions will require an ability to function effectively on interdisciplinary teams of individuals with very different backgrounds. In addition, chemical engineers will need a greater appreciation of economic constraints, policy and regulatory issues, and public perceptions.

What do you enjoy most about your job? What do you enjoy least about your job?
Chemical engineering is a vibrant field that builds directly on developments in the basic sciences and helps convert those discoveries into solutions to major societal challenges. To me, research and the challenge of working on a problem that no one has ever solved before is incredibly exciting. Teaching can be hugely satisfying as well; it's wonderful to see a student have that "a ha" moment when everything falls into place after having struggled with a complex concept.

To be successful in research, one must write proposals that will generate external funding to support a research program. It's time-consuming and often frustrating; competition is high and most grant proposals are rejected-so that's probably the least enjoyable part of my job.

Can you suggest a valuable "try this" for students considering a career in your profession?
Playing with a good chemistry set is a great way to see if you enjoy chemistry. Do number puzzles and brainteasers to see if you enjoy solving problems because chemical engineers must be excellent problem solvers. Entering a science fair will give you a feel for what research is all about. As for teaching, help other students with difficult problems or concepts to see if you have a natural ability to teach and communicate.

This interview was originally published in 2015.

Civil Engineer

Snapshot

Career Cluster(s): Architecture & Construction; Science, Technology, Engineering & Mathematics; Transportation, Distribution & Logistics

Interests: Construction; Design; Mathematics; Problem-solving; Research

Earnings (Yearly Average): $88,570

Employment & Outlook: As Fast As Average Growth Expected

OVERVIEW

Sphere of Work

Civil engineers conceive, design, build, supervise, operate, construct, and maintain infrastructure projects and systems in the public and private sector, including roads, buildings, airports, tunnels, dams, bridges, and systems for water supply and sewage treatment. Many civil engineers work in planning, design, construction, research, and education.

Work Environment

Civil engineers work in a variety of locations and conditions. When working on designs, civil engineers may spend most of their time indoors in offices. However, construction engineers may spend much of their time outdoors at construction sites monitoring operations or solving on-site problems. Some jobs may require frequent relocation to different areas and offices in jobsite trailers.

Photo via iStock/sculpies. [Used under license.]

Occupation Interest

Civil engineering satisfies both the creative and analytical mind. This profession combines design concepts with mathematics and science to create infrastructure that is used on a daily basis, by either the general public or public and private sector workers. Without civil engineers, we wouldn't have the advanced and developed communities or systems we see today. Because civil engineers have to ensure that the structures they create are sound, attention to detail is imperative in this field. Someone with a keen eye and the ability to work well in pressured situations will excel in a civil engineering position. Overall, these professionals have a desire to not only maintain infrastructure for public safety, but find way to improve it as well.

Profile

Working Conditions: Both Inside and Outside
Physical Strength: Light Work; Medium Work
Education Needs: Bachelor's Degree
Licensure/Certification: Optional
Opportunities for Experience: Internship, Apprenticeship, On-the-Job Training
Interest Score: RIC

A Day in the Life—Duties and Responsibilities

Civil engineers design, build, and maintain various infrastructures. They are responsible for the roadways, railways, airports, bridges, and much more that we use on a daily basis. The work of civil engineers is closely related to the work of environmental engineers.

Civil engineers must present their findings to the public on topics such as bid proposals, environmental impact statements, or property descriptions.

Many civil engineers hold supervisory or administrative positions ranging from supervisor of a construction site to city engineer, public works director, and city manager. As supervisors, they are tasked with ensuring that safe work practices are followed at construction sites.

Other civil engineers work in design, construction, research, and teaching. Civil engineers work with others on projects and may be assisted by civil engineering technicians.

Duties and Responsibilities

- Analyzing long-range plans, surveying reports, maps, and other data to plan and design projects
- Understanding construction costs, government regulations, potential environmental hazards, and other factors during the planning and risk-analysis stages of a project
- Compiling and submitting permit applications to local, state, and federal agencies, verifying projects comply with various regulations
- Overseeing and analyzing the results of soil testing to determine the adequacy and strength of foundations
- Analyzing the results of tests on building materials, such as concrete, wood, asphalt, or steel, for use in particular projects
- Preparing cost estimates for materials, equipment, or labor to determine a project's economic feasibility
- Using design software to plan and design transportation systems, hydraulic systems, and structures in line with industry and government standards

Civil engineers prepare permit documents for work on projects in renewable energy. They verify that the projects will comply with federal, state, and local requirements. These engineers conduct structural analyses for large-scale photovoltaic, or solar energy, projects. They also evaluate the ability of solar array support structures and buildings to tolerate stresses from wind, seismic activity, and other sources. For large-scale wind projects, civil engineers often prepare roadbeds to handle large trucks that haul in the turbines.

Civil engineers work on complex projects, and they can achieve job satisfaction in seeing the project reach completion. They usually specialize in one of several areas.

OCCUPATION SPECIALTIES

Construction Engineer
Construction engineers manage construction projects, ensuring that they are scheduled and built in accordance with plans and specifications. These engineers typically are responsible for the design and safety of temporary structures used during construction. They may also oversee budgetary, time-management, and communications aspects of a project.

Geotechnical Engineer
Geotechnical engineers work to make sure that foundations for built objects ranging from streets and buildings to runways and dams are solid. They focus on how structures built by civil engineers, such as buildings and tunnels, interact with the earth (including soil and rock). In addition, they design and plan for slopes, retaining walls, and tunnels.

Structural Engineer
Structural engineers design and assess major projects, such as buildings, bridges, or dams, to ensure their strength and durability.

Transportation Engineer
Transportation engineers plan, design, operate, and maintain everyday systems, such as streets and highways, but they also plan larger projects, such as airports, ship ports, mass transit systems, and harbors.

WORK ENVIRONMENT

Immediate Physical Environment

Although civil engineers must work in an office setting to produce their plans and meet with clients or government officials, they must also spend time on-site to oversee construction. Before a concept is translated into a real-world project, they will work on the planning and designing stages indoors. When a project moves into a building and construction stage, engineers may be expected to oversee operations on-site and use their problem-solving skills to provide solutions for potential roadblocks. Depending on the job a civil engineer is working on, they may be required to relocate to different areas often and work out of office trailers.

Civil engineers who function as project managers may work from cars or trucks as they move from site to site. Many civil engineers work for government agencies in government office buildings or facilities. Occasionally, civil engineers travel abroad to work on large engineering projects in other countries.

Human Environment

Civil engineers typically communicate with a wide array of people on a daily basis, from clients and government officials to construction crews, architects, material suppliers, and more. In order to ensure that the plan they've conceived is understood by all parties involved and being executed correctly, they must be able to explain their designs and findings in a way that can be easily understood by people with little or no engineering and scientific background. Civil engineers may also work closely with other engineers on a project, brainstorming with peers to develop plans or provide solutions to problems.

Technological Environment

Civil engineers use a variety of technology and tools to complete their work. The computer software they utilize ranges from computer-aided design (CAD) and analytical and scientific software to Office suite software. In addition, they may use geographic information systems and cartography software to map out the site for construction. When they are attending a project site, they'll need to know how to use tools for surveying the construction area. These could include levels, compasses, tools for measuring land and water area, tools for testing water and soil quality, and more. Depending on the type of project, a civil engineer may need to know how to operate construction machinery.

EDUCATION, TRAINING, AND ADVANCEMENT

High School/Secondary

High school students interested in a career in civil engineering will benefit from taking courses in math and science. These two subjects are essential to the role of a civil engineer and are utilized daily. Taking English and public speaking courses are very valuable for sharpening communication and interpersonal skills. Because civil engineers use many different types of software, courses in computer science or computer programming may give a student a jumpstart on understanding these complex systems. In addition, taking art classes can help an aspiring civil engineer with designing infrastructure and creating blueprints in the future.

Suggested High School Subjects

- Algebra
- Biology
- Calculus
- Chemistry
- Civics
- Computer Programming
- Earth or Life or Physical Science
- Economics
- English
- Entrepreneurship
- Geometry
- History
- Math Analysis
- Physics
- Political Science
- Psychology
- Statistics
- Trigonometry
- World Geography

Related Career Pathways/Majors

Architecture & Construction Career Cluster

- Design/Pre-Construction Career Pathway

Science, Technology, Engineering & Mathematics Career Cluster

- Engineering & Technology Career Pathway

Transportation, Distribution & Logistics Career Cluster

- Transportation Systems/Infrastructure Planning, Management & Regulation Career Pathway

Postsecondary

Civil engineers typically need a bachelor's degree in civil engineering, civil engineering technologies, or a related field, such as construction. Programs in civil engineering and civil engineering technology include coursework in math, statistics, engineering mechanics and systems, and fluid dynamics, depending on the specialty. Courses include a mix of traditional classroom learning, work in laboratories, and fieldwork. Programs may include co-operative programs, also known as co-ops, in which students gain work experience while pursuing a degree.

A degree from a program accredited by the Accreditation Board for Engineering and Technology, Inc. (ABET) is needed to earn the professional engineer (PE) license. In many states, a bachelor's degree in civil engineering technology also meets the academic requirement for obtaining a license.

Related College Majors

- Architectural Engineering
- Architecture
- Civil Engineering
- Drafting and Design Technology
- Engineering Technology
- Environmental Engineering

Transferable Skills and Abilities

Decision-making Skills

- Balancing multiple and frequently conflicting objectives
- Making good decisions based on best practices, technical knowledge, and personal experience

Leadership Skills

- Taking ultimate responsibility for the projects they manage or research they perform
- Leading planners, surveyors, construction managers, civil engineering technicians, civil engineering technologists, and others in implementing their project plan

Math Skills

- Using the principles of calculus, trigonometry, and other advanced topics for analysis, design, and troubleshooting

Organizational Skills

- Monitoring and evaluating the office work and the jobsite as the project progresses
- Ensuring compliance with design documents
- Managing several projects at the same time
- Balancing time needs and effectively allocating resources

Problem-solving Skills

- Identifying and evaluating complex problems
- Using their skill and training to develop cost-effective, safe, and efficient solutions

Speaking Skills

- Presenting reports and plans to audiences of people with a wide range of background and technical knowledge
- Speaking clearly and conversing with people in various settings
- Translating engineering and scientific information into easy-to-understand concepts

Writing Skills

- Communicating with others, such as architects, landscape architects, urban and regional planners
- Explaining projects to elected official and citizens
- Writing reports that are clear, concise, and understandable to those with little or no technical or scientific background

Adult Job Seekers

Ideally, adults seeking a career in civil engineering have experience in a similar or related field. A bachelor's degree in a relevant subject is necessary work in this profession. Adult job seekers without a related degree or applicable experience that are interested in becoming a civil engineer need to enroll in relevant continuing education courses or in a postsecondary accredited college or university. Without receiving a bachelor's degree in civil engineering or a similar subject, there are no opportunities for an adult to work in this field.

Adults that are already qualified civil engineers may find work through a variety of sources, including the government, engineering and architectural firms, or professional associations. Networking through engineering societies may provide both new and seasoned civil engineers with broad opportunities.

Professional Certification and Licensure

Licensure is not required for entry-level positions as a civil engineer. A Professional Engineering (PE) license, which allows for higher levels of leadership and independence, can be acquired later in one's career. Licensed engineers are called professional engineers (PEs). A PE can oversee the work of other engineers, approve design plans, sign off on projects, and provide services directly to the public. State licensure generally requires a degree from an ABET-accredited engineering program; a passing score on the Fundamentals of Engineering (FE) exam; relevant work experience, typically at least 4 years working under a licensed engineer; and a passing score on the Professional Engineering (PE) exam.

The initial FE exam can be taken after earning a bachelor's degree. Engineers who pass this exam commonly are called engineers in training (EITs) or engineer interns (EIs). After meeting work experience requirements, EITs and EIs can take the second exam, called the Principles and Practice of Engineering.

Each state issues its own licenses. Most states recognize licensure from other states, as long as the licensing state's requirements meet or exceed their own licensure requirements. Several states require continuing education for engineers to keep their licenses.

The American Society of Civil Engineers (ASCE) offers certifications in coastal engineering, geotechnical engineering, ports engineering, water resources engineering, and other fields. Additionally, civil engineers can become certified in building security and in sustainability.

Additional Requirements

Civil engineers, in addition to having an advanced understanding of math and science, must be able to use their knowledge and expertise to provide creative solutions to a variety of problems. Because this profession requires strong communication skills, civil engineers also need to be able to work in a team setting. At times, civil engineers may need to step into a leadership position to supervise projects and give direction.

EARNINGS AND ADVANCEMENT

Earnings of civil engineers depend on their education and experience, employer, licensure, specialty, and the types of projects on which they work. Median annual earnings of civil engineers were $88,570 in 2020. The lowest 10 percent earned less than $56,160, and the highest 10 percent earned more than $144,810.

Civil engineers may receive benefits such as paid vacations, holidays, and sick days; life and health insurance; and retirement benefits, usually paid by the employer.

Civil engineers with ample experience may move into senior positions, such as project managers or functional managers of design, construction, operation, or maintenance. However, they would first need to obtain the Professional Engineering (PE) license, because only licensed engineers can assume responsibilities for public projects.

After gaining licensure, a professional engineer may seek credentialing that demonstrates his or her expertise in a civil engineering specialty. Such a credential may be helpful for advancement to senior technical or even managerial positions.

EMPLOYMENT AND OUTLOOK

There were approximately 309,800 civil engineers employed nationally in 2020. Employment of civil engineers is expected to grow as fast as average for all occupations through the year 2030, at a rate of 8 percent.

As current U.S. infrastructure experiences growing obsolescence, civil engineers will be needed to manage projects to rebuild, repair, and upgrade bridges, roads, levees, dams, airports, buildings, and other structures.

Famous First

The world's first Ferris Wheel was created by civil engineer George W. Ferris, for the World's Columbian Exposition in Chicago, in 1893. The fair's creators wanted something that would rival the Eiffel Tower, and at first didn't believe the towering wheel would be safe. Source: ie-services.com

Photo via iStock/ilbusca. [Used under license.]

A growing population likely means that new water systems will be required while, at the same time, aging, existing water systems must be maintained to reduce or eliminate leaks. In addition, more waste treatment plants will be needed to help clean the nation's waterways. Civil engineers will continue to play a key part in all of this work.

The work of civil engineers will be needed for renewable-energy projects. Thus, as these new projects gain approval, civil engineers will be further involved in overseeing the construction of structures such as wind farms and solar arrays.

Although state and local governments continue to face financial challenges and may have difficulty funding all projects, some delayed projects will have to be completed to build and maintain critical infrastructure, as well as to protect the public and the environment.

Related Occupations

- Architect
- Civil Engineering Technologist/Technician
- Construction Manager
- Environmental Engineer
- Landscape Architect
- Mechanical Engineer
- Surveyor
- Urban/Regional Planner

MORE INFORMATION

Accreditation Board for Engineering and Technology, Inc. (ABET)
415 N. Charles Street
Baltimore, MD 21201
410.347.7700
comms@abet.org
www.abet.org

American Society for Engineering Education (ASEE)
1818 N Street, NW, Suite 600
Washington, DC 20036
202.331.3500
www.asee.org

American Society of Civil Engineers (ASCE)
1801 Alexander Bell Drive
Reston, VA 20191
800.548.2723
www.asce.org

National Council of Examiners for Engineering and Surveying (NCEES)
200 Verdae Boulevard
Greenville, SC 29607
800.250.3196
www.ncees.org

National Society of Professional Engineers (NSPE)
1420 King Street
Alexandria, VA 22314
888.285.6773
www.nspe.org

Technology Student Association (TSA)
1904 Association Drive
Reston, VA 20191-1540
703.860.9000
general@tsaweb.org
www.tsaweb.org

Emily Porter

Conversation With...

MICHAEL RAMSEY, P.O.

Director of Public Works, Village of Westmont
Westmont, IL
Years in public works, 34

What was your individual career path in terms of education/training, entry-level job, or other significant opportunity?

I moved around 18 times in 18 years growing up because my dad was a chemical engineer in the oil and gas business. Then we settled in Illinois. I went to college thinking I wanted to be an accountant and took a job with the Village of Westmont's Public Works Department as a seasonal employee. Due to financial issues, I ended up dropping out of college. The village offered me a fulltime position and I jumped at it. I started as a laborer and went on to be a foreman, equipment operator, water operator, water superintendent, and now I'm the Director of Public Works.

A few years into my career, I went back to Illinois State University and finished my business administration degree online.

It's been a journey, and it's been fun. I've gotten to know a lot of people in the community. I've been involved with the American Water Works Association at the local, state and national level. That includes sitting on the executive board at the state level, and creating a competition for the annual conference where two-person teams of water operators from around the world compete to see how fast they can put a water hydrant together. It's called Hydrant Hysteria, and helps operators get more involved with the association and builds camaraderie.

I also established the Operator in Training program in our local high school. Students can learn how to become a Certified Class D Water Treatment Operator in the state of Illinois through a 64-hour online course. I go to the school on Fridays to instruct and one Saturday a month the students come into the water treatment plant for hands-on experience. I show them how to fix a water main break, install a water meter—all of it. In summer I hire them as seasonal employees or interns. It's a good opportunity for students who don't know what they want to do yet, or to see if they'd like to do this. I'm planning to hire one of my students full-time when our current hiring freeze ends.

What are the most important skills and/or qualities for someone in your profession?
Being somewhat strong in STEM (science, math, engineering, technology) skills really helps. If you're a water operator, there's a lot of chemistry and math every day. If you're pouring concrete, you need to figure out how much concrete to pour.

You need to be willing to work outside, sometimes in the cold, sometimes in the heat. Sometimes it's beautiful. You need to be physically fit, able to lift 50-75 pounds. You don't need to be an Olympic athlete. You have to be somewhat outgoing because you're dealing with local residents, media and politicians.

What do you wish you had known going into this profession?
I regret not getting involved sooner with professional organizations. Do it as early as possible because it opens a lot of doors.

Are there many job opportunities in your profession? In what specific areas?
I'm always looking for good men and women. We need water operators, equipment operators. Baby boomers are retiring and there are a lot of job openings with great benefits and a pension. This is a really great opportunity for career advancement.

How do you see your profession changing in the next five years, how will technology impact that change, and what skills will be required?
Everything is automated. I can operate our treatment plant from my phone. Keeping up with changing technology will continue.

However, our country's infrastructure is really starting to fail. That's what's driving this industry. Water mains were put in 100 years ago. Who's going to fix and replace that? We need to be making sure our roads and bridges are good. Public works is going to come to the forefront because this stuff is starting to fail and if we don't take care of it, we're going to be in big trouble.

What do you enjoy most about your job? What do you enjoy least about your job?
I most like the science part of my job, and making a difference in my town and giving back to our residents. Every day I learn something new, including from both the kids and adults I teach.

I don't like dealing with customers who are irate. You have to learn to be patient, put yourself in their position, and also how to take a breath and leave it at work. Sometimes there is no way to fix their problem, but it's hard to tell them that when their basement is flooding and they need to get to work.

Can you suggest a valuable "try this" for students considering a career in your profession?
There are a lot of online courses in public works, and some of it, like webinars, are free. Especially for students. Check the American Water Works Association at awwa.org or its Illinois section at isawwa.org.

This interview was originally published in 2020.

Conversation With...

JAMES W. BLAKE, P.E., P.L.S.

Owner
Blake Consulting Services, LLC
Civil Engineer, 35 years

What was your individual career path in terms of education/training, entry-level job, or other significant opportunity?

My father was a civil engineer, and although I wasn't quite sure what I wanted to do, I knew if I had that degree, I'd have options. I earned a B.S. in civil engineering from the University of Maryland, and considered becoming a pilot. Unfortunately, my vision wasn't good enough. So I went to work as a civil/structural engineer, then went right back to school and earned a B.S. in business and accounting because I was interested in management and the business of engineering. My first job was a lot of sitting at a desk and crunching numbers and I wanted more variety, so I moved into transportation and general civil engineering, which allowed me to interact with clients, politicians and a greater variety of projects. I got into management about eight to 10 years out of college. My specialty is roadway design and construction management. A lot of what I do now is business development related and involves competing for and winning consulting contracts from federal, state, and local governmental agencies.

I obtained my LEED Accredited professional certification in 2007 which, at the time, was the leading edge for environmental practice within civil engineering. I foresaw the future emphasis on conservation and green development, and that certified engineers would fill a big need. LEED, which stands for Leadership in Energy and Environmental Design, is the system used by the U.S. Green Building Council to measure a building's sustainability and efficiency

What are the most important skills and/or qualities for someone in your profession?

The ability to communicate technical ideas in plain, easily understood language is critical. You also need to be organized, and to have the ability to analyze a complex problem quickly, break it down into its various parts and then come up with a practical and economical solution that serves the client's best interests.

What do you wish you had known going into this profession?

Construction and civil engineering can be cyclical due to economic cycles. So, you have to be able to adapt and change. The niches within the profession come and go, and your skills must be diverse enough to shift where you're needed. The politics of winning work through public agencies is also more involved and complex than I'd imagined it to be. Someone who wants to win public sector work must be very visible and involved with a particular governmental agency and those who oversee it.

Are there many job opportunities in your profession? In what specific areas?
I'd say very many. Job opportunities are excellent in transportation, structural, civil, environmental and geotechnical engineering, as well as construction inspection. Civil engineering is a very broad and diverse field, and there are some great challenges and opportunities.

In terms of green infrastructure, opportunity lies within water resources, where you deal with drainage, storm water management, best management practices for water quality, and stream restoration. The water and sewer treatment sector also offers many opportunities.

LEED accredited professionals tend to deal with buildings, but civil engineers will find additional opportunity if they are certified in a similar green standard for roads, called the Envision system (ENV-SP). So, maybe you would use specialized materials that are more environmentally friendly, or recycle broken-up concrete and use that as a subbase for a road.

How do you see your profession changing in the next five years, what role will technology play in those changes, and what skills will be required?
Technology continues to radically change the way things get done in my profession. Only a few years ago, CAD drawings were big. Now, 3D Building Information Modelling (BIM) technology is the standard. It allows you to have a 3D model of a structure and you can cut a plane through it to produce plans. All kinds of new software programs are coming along to do various analytical tasks, and that will continue. For instance, you used to do studies to develop a flood plain. Now, the results are produced visually, then superimposed on a model of a city so you can see the depth of flooding in different areas. New technology also allows laser scanning in the field.

What do you enjoy most about your job? What do you enjoy least about your job?
I most enjoy the variety of projects and my ability to interact with various people who have different roles in the development and completion of a project. Civil engineering has more of a people side to it than other types of engineering. I least enjoy times of economic duress, which all businesses undergo at some point, and the downsizing that results. I've had to lay people off in my career, and I dislike doing it.

Can you suggest a valuable "try this" for students considering a career in your profession?
Find an engineering company and see if they sponsor a student mentor day. Some professional organizations do that, such as the American Society of Civil Engineers or the Society of American Military Engineers. Also, there are parks that have storm water management and wetland exhibits built right in to them for public education. Find one of those and you can see an example of a green project by a civil engineer.

This interview was originally published in 2015.

Conversation With...
LISA DEITEMEYER
Project Manager

What was your career path?

I graduated from Northeastern University with a Bachelor of Science degree in civil engineering. While still in college I passed the Fundamentals of Engineering exam. With this I became an Engineer in Training (EIT). After graduation I worked for a transportation engineering consulting firm. As an EIT you need to work five years (varies in different states) under the supervision of a professional engineer before you can take the professional engineers (PE) exam. My experience working as an EIT was primarily in roadway and drainage design for construction projects varying in size from residential streets to highway interchanges. With good engineering design experience on progressively complex work, I took the PE exam, passed the exam, and officially became a professional engineer. I continued to work as a designer and project engineer on transportation projects. With about nine years of experience, the company I worked for provided the opportunity for me to manage the design of a residential street reconstruction project. From there I continued to have opportunities to manage larger projects. With my increase in management work, my design work decreased and I supervised the design work of others. Today, I am a project manager for the design of a $2 billion design-build construction project in Dallas, Texas.

What are three pieces of advice?

1. Try to work as an intern or get a summer job at a company in the field you are studying. This exposure to real life work will help you confirm that it is the field for you, and it will help apply your college course work to the working world.

2. As and EIT or graduate engineer, try to get as much experience as you can in the areas of your field. For example, in civil engineering, try to get experience in the design of drainage, roadway, traffic, retaining walls, etc. Understanding all aspects of your field will help you become a better manager later, if you choose that path.

3. Learn as much as you can from others that are successful in your field. Observe your successful managers and how they interact with their staff and clients. Ask questions and have a clear understanding of your career advancement path and professional goals.

What paths for career advancement are available to you?

At this point in my career, advancement that is available to me includes the pursuit of larger projects which includes doing the marketing to pursue new business, preparing proposals for new project, and then managing the projects if we are successful in winning them. Other career advancement that is available to me is to obtain a higher position within the firm I work for such as a vice president. With this advancement I could become a decision maker for the company as a whole.

This interview was originally published in 2013.

Civil Engineering Technologist/ Technician

Snapshot

Career Cluster(s): Architecture & Construction; Manufacturing; Science, Technology, Engineering & Mathematics; Transportation, Distribution & Logistics

Interests: Analyzing Data; Construction; Mathematics; Problem-solving

Earnings (Yearly Average): $54,080

Employment & Outlook: Slower Than Average Growth Expected

OVERVIEW

Sphere of Work

Civil engineering technologists and technicians help civil engineers plan, design, and build highways, bridges, utilities, and other infrastructure projects. They also help to plan, design, and build commercial, industrial, residential, and land development projects.

Photo via iStock/aeduard. [Used under license.]

Work Environment

Civil engineering technologists and technicians work in offices, helping civil engineers plan and design projects. They also visit construction jobsites to collect or test materials or to observe activities as a project inspector. They may work at several sites, using cars or trucks as a mobile office.

Profile

Working Conditions: Both Inside and Outside
Physical Strength: Light Work; Medium Work
Education Needs: Associate's Degree
Licensure/Certification: Not Required
Opportunities for Experience: Internship, On-the-Job Training
Interest Score: RCI

Occupation Interest

Civil engineering technologists and technicians are typically very methodical in their work and can be found working out the logistics and technical side of a project. They may have strong interests in math and science, particularly when applying the principles of each to real-world projects. They are the team members on a project that help the head engineers take their design from a concept to a full-fledged idea, seeing it through to construction. Civil engineering technologists and technicians can find fulfillment through problem-solving and collecting data.

A Day in the Life—Duties and Responsibilities

Civil engineering technicians typically install, troubleshoot, and maintain designs created by engineers. They may work under the direction of engineers or engineering technologists.

Duties and Responsibilities

- Reading and reviewing project drawings and plans to determine the sizes of structures
- Conferring with engineers to prepare plans
- Developing plans and estimating costs for constructing systems and operating facilities
- Using computer-aided design (CAD) software to draft project drawings
- Conducting field surveys to collect data on site conditions and inspecting structures
- Testing appropriateness of constructing materials and soil samples
- Observing project sites to ensure that construction conforms to design specifications and applicable codes
- Preparing reports and documents project activities and data
- Setting up and helping maintain project files and records

Civil engineering technologists typically help licensed engineers improve designs or incorporate new technology. They may be team leaders, instructing civil engineering technicians on installing equipment, systems, or structures.

These technologists and technicians observe progress on a jobsite, collect data, and complete reports to document project activities. Because they are not licensed, civil engineering technologists and technicians cannot approve designs or supervise the overall project.

In addition, civil engineering technologists and technicians sometimes estimate construction costs, develop specifications, and prepare drawings. They also may set up and monitor various instruments for traffic studies. Their duties often require use of software to design projects, collect and analyze data, prepare reports, and manage files.

WORK ENVIRONMENT

Immediate Physical Environment

Civil engineering technologists and technicians complete work both indoors and outdoors. In an office setting, they help civil engineers with the conceptualization of their projects. This ranges from drawing up designs with CAD software to calculating costs and prepping reports.

While their indoor work involves mostly paperwork and preparing plans, outdoor work consists of attending constructions sites for observation and documentation of a project's progress. If a project isn't conforming to the design's parameters, or if the construction material or foundation is low quality, the civil engineering technologist records their observations to report back to their supervisor.

Civil engineering technologists and technicians frequently work on teams with civil engineers, surveyors and surveying technicians, construction workers, and others involved with projects.

Human Environment

The projects that civil engineering technologists and technicians work on require close communication with a team of professionals. Without clear communication, the infrastructure being built could end up being deemed unsound, which in turn would be much more costly to fix. Civil engineering technicians need to be able to report their findings to civil engineers and work collaboratively with others to ensure a project runs smoothly from its conception to its execution.

Technological Environment

Civil engineering technologists and technicians utilize many different types of computer software and technology on a daily basis. The software they use includes CAD software, analytical and scientific software, graphics and photo imaging software, and office suite software. In addition, they may use geographic information systems and map creation software for planning out a construction site. When a civil engineering technologist is attending a project site, they'll use tools for surveying the construction area. Particularly, they'll use tools to test the quality of materials, water, and soil.

EDUCATION, TRAINING, AND ADVANCEMENT

High School/Secondary

High school students that enjoy courses in math and science may find fulfillment through a career as a civil engineering technologist or technician. Taking advanced placement courses in these subjects when available will better prepare someone interested in this profession. To improve communication skills—verbal and written—aspiring civil engineering technicians should take English classes. If a student's school offers courses in computer programming or computer science, these courses may help a student hone their skills early on, before applying them to the computer software specific to this profession.

Suggested High School Subjects

- Algebra
- Biology
- Calculus
- Chemistry
- Civics
- Computer Programming
- Earth or Life or Physical Science
- Economics
- English
- Entrepreneurship
- Geometry
- History
- Math Analysis
- Physics
- Psychology
- Statistics
- Trigonometry
- World Geography

Transferable Skills and Abilities

Critical-thinking Skills
- Carrying out project plans and designs that engineers have approved
- Understanding and interpreting the reports and documents describing these projects

Decision-making Skills
- Discerning which information is most important and which actions will help keep a project on schedule

Math Skills
- Using math for analysis, design, and troubleshooting

Observational Skills
- Visiting jobsites to assess a project and reporting back to the engineer

Problem-solving Skills
- Helping engineers identify problems and designing projects to solve them

Writing Skills
- Preparing reports that are well-organized and clearly convey information

Related Career Pathways/Majors

Architecture & Construction Career Cluster
- Design/Pre-Construction Career Pathway

Manufacturing Career Cluster
- Manufacturing Production Process Development Pathway

Science, Technology, Engineering & Mathematics Career Cluster
- Engineering & Technology Career Pathway

Transportation, Distribution & Logistics Career Cluster

- Transportation Systems/Infrastructure Planning, Management & Regulation Career Pathway

Postsecondary

An associate's degree, preferably in civil engineering technology, is typically required to enter the occupation. Prospective civil engineering technologists and technicians should take courses in science and math, such as chemistry, physics, geometry, and trigonometry.

Employers may prefer to hire engineering technologists and technicians who have an associate's degree from a program accredited by the Accreditation Board for Engineering and Technology, Inc. (ABET), although a degree is not always required. Engineering technology programs also are available at technical or vocational schools that award a postgraduate certificate or diploma.

Degree and nondegree programs may include coursework in subjects such as engineering, design, and sciences.

Some employers require a bachelor's degree in engineering technology for civil engineering technologists.

Related College Majors

- Architecture
- Civil Engineering
- Civil Engineering Technology
- Engineering Technology

Adult Job Seekers

Adults interested in working as a civil engineering technologist or technician should enroll in either an accredited associate's degree program or a postgraduate certification program. Although an associate's degree may be preferred by some employers, it is not always required. Many technical and vocational schools offer certificates and diplomas in civil engineering technology, both of which are valid credentials when searching for work in this field. Previous experience in a similar or related field will benefit an adult seeking a career in this profession.

Professional Certification and Licensure

Certification is not required to enter this occupation, but it may help technologists and technicians develop in their careers. For example, the National Institute for

Certification in Engineering Technologies (NICET) oversees certification for civil engineering technicians who pass an exam and provide supporting documentation. NICET requires technicians to update their skills and knowledge through a recertification process that encourages continuing professional development.

Additional Requirements

Civil engineering technicians should be comfortable working with a team on projects and be able to communicate effectively. They typically have an interest in construction and infrastructure design and a desire to work both indoors and outdoors. Civil engineering technicians and technologists are able to collaboratively brainstorm solutions to a project's problems by working directly with lead civil engineers.

EARNINGS AND ADVANCEMENT

Earnings of civil engineering technologists and technicians depend on their experience, employer, and the types of projects on which they work. Median annual earnings of civil engineering technologists and technicians were $54,080 in 2020. The lowest 10 percent earned less than $34,230, and the highest 10 percent earned more than $82,960.

Civil engineering technicians may receive benefits such as paid vacations, holidays, and sick days; life and health insurance; and retirement benefits, usually paid by the employer.

Civil engineering technologists and technicians may advance in their careers by learning to design systems for a variety of projects, such as storm sewers, and to become skilled at reading graphical drawings of proposed projects.

Technicians who obtain appropriate education or certification may advance to become technologists.

EMPLOYMENT AND OUTLOOK

There were approximately 68,800 civil engineering technicians and technologists employed nationally in 2020. Employment of civil engineers is expected to grow slower than the average for all occupations through the year 2030, at a rate of 2 percent.

The need to preserve, repair, and upgrade the country's infrastructure continues to increase. Bridges, roads, levees, airports, and dams will need to be rebuilt, maintained, and enhanced. Also, a growing population means that water systems must be maintained in order to reduce or eliminate loss of drinkable water. In addition, more waste treatment plants will be needed to help clean the nation's waterways. Civil engineers must plan, design, and oversee this work, and civil engineering technologists and technicians will be needed to assist the engineers in these projects.

Civil engineering technologists and technicians also will find work assisting civil engineers with renewable-energy projects. For example, these technologists and technicians may assist in the development of a wind farm by helping engineers devise solutions to minimize project costs while accommodating the sizeable dimensions and weight of wind turbines.

Related Occupations

- Civil Engineer
- Drafter
- Surveying/Mapping Technician
- Surveyor

MORE INFORMATION

Accreditation Board for Engineering and Technology, Inc. (ABET)
415 N. Charles Street
Baltimore, MD 21201
410.347.7700
comms@abet.org
www.abet.org

National Institute for Certification in Engineering Technologies (NICET)
1420 King Street
Alexandria, VA 22314-2794
888.476.4238
www.nicet.org

Pathways to Science
P.O. Box 607
Damariscotta, ME 04543
207.424.1185
contactus@ibparticipation.org
www.pathwaystoscience.org

Emily Porter

Computer Hardware Engineer

Snapshot

Career Cluster(s): Information Technology; Science, Technology, Engineering & Mathematics

Interests: Computers; Computer Science; Problem-Solving; Science

Earnings (Yearly Average): $119,560

Employment & Outlook: Slower Than Average

OVERVIEW

Sphere of Work

Computer hardware engineers research, design, develop, and test computer systems and components such as processors, circuit boards, memory devices, networks, and routers. Many hardware engineers design devices used in manufactured products that incorporate processors and other computer components and that connect to the internet. For example, many new cars, home appliances, and medical devices have internet-ready computer systems built into them. Computer hardware engineers ensure that computer hardware components work together with the latest software. Therefore, hardware engineers often work with software developers. For example, the hardware and software for mobile phones and other devices frequently are developed at the same time.

Photo via iStock/scanrail. [Used under license.]

Work Environment

Most computer hardware engineers typically work inside research laboratories, offices, or manufacturing firms where they test different types of computer models. The location of these research labs is most commonly found in or nearby large metropolitan cities. The majority of a computer hardware engineer's time would be spent working at their computer workstation. Engineers may also spend their time overseeing the installation and testing of what they created. In these instances, proper safety precautions must be taken—including a sterile environment—to ensure the safety of each engineer. Most computer hardware engineers work a standard full-time week of 40 hours and depending the time sensitivity of the workload, overtime may be required to complete projects.

Profile

Working Conditions: Mainly Inside
Physical Strength: Light Work
Education Needs: Bachelor's Degree, Master's Degree
Licensure/Certification: Typically Not Required
Opportunities for Experience: Internship
Interest Score: IRC

Occupation Interest

A career as a computer hardware engineer will appeal to individuals who not only have a strong interest in computer science and engineering, but those who are strong critical thinkers when faced with complex problems. Computer hardware engineers must troubleshoot and find solutions for hardware and computer problems when it comes to their work. They should also be interested in the latest technology and open to constantly learning as technology continues to advance.

A Day in the Life—Duties and Responsibilities

Computer hardware engineers use their education and skills to design, develop, and test new computer hardware and components. These include computer systems, computer chips, and the physical parts of computers. Engineers are also involved in the design and development of new routers, printers, and keyboards. Their daily responsibilities vary depending on the project they are working on. Meetings may be held throughout the day with other engineers, technology vendors, and various other employees.

Duties and Responsibilities

- Designing new computer hardware and creating schematics of computer equipment to be built
- Testing the computer hardware they design
- Analyzing the test result and modifying the design as needed
- Updating existing computer equipment so that it will work with new software
- Overseeing the manufacturing process for computer hardware

Computer hardware engineers are normally involved in the entire process of product development and implementation. This includes the manufacturing process. Throughout the day, an engineer will provide technical support to other employees, including designers, the marketing department, and technology vendors. As computer technology is developed and created, engineers will perform tests and ensure that everything meets specifications and requirements. This testing process usually involves analyzing test data, product prototypes, or theoretical models. As new hardware is implemented, an engineer will monitor how it is functioning and make any modifications necessary so that it performs according to specifications. Engineers also make recommendations for additional hardware, such as keyboards, routers, and printers.

When new components are designed and manufactured, engineers have to make sure that the hardware is compatible with software developments. Because of this, hardware engineers collaborate closely with software developers throughout the process.

WORK ENVIRONMENT

Immediate Physical Environment

The physical work environment for computer hardware engineers is predominantly in a research lab. These spaces are where they would design, build, and test what they have built. Overall, the laboratories where engineers work are typically bright and open spaces that are clean and well-ventilated.

Human Environment

Throughout the day, computer hardware engineers can go between solitarily working on projects and communicating with their colleagues when needed. Engineers may also find themselves collaborating with other professionals in similar fields, ranging from software engineers, manufacturers, other engineers, or technology vendors.

Technological Environment

Computer hardware engineers work with a wide array of computer-based technologies that range from circuit boards and processors, to chips, electronic equipment, and computer hardware and software, including applications and programming. During the design process, computer-aided design (CAD) software is used, as well as servers that store large amounts of data.

EDUCATION, TRAINING, AND ADVANCEMENT

High School/Secondary

A requirement of employers is that an applicant must have a high school diploma or a GED certificate equivalent. High school courses such as science, mathematics, engineering, or computer science will greatly benefit anyone interested in pursuing a career as a computer hardware engineer. Schools that offer extracurricular clubs involving computer science, robotics, or science are also beneficial to those wishing to get into engineering.

Suggested High School Subjects

- Algebra
- Applied Physics
- Biology
- Calculus
- Chemistry
- Computer Programming
- Computer Science
- English
- Geometry
- Microbiology
- Organic Chemistry
- Structured Computer Program Language
- Trigonometry

Transferable Skills and Abilities

Analytical Skills
- Using tools to analyze the digital circuits in hardware and determining the best design

Critical-thinking Skills
- Using logic and reasoning to clarify goals
- Examining assumptions
- Identifying the strengths and weaknesses of alternative solutions

Problem-solving Skills
- Indentifying complex problems in computer hardware
- Developing and evaluating possible solutions
- Figuring out the best way to implement solutions

Communication Skills
- Communicating clearly and efficiently with other engineers, software developers and programmers, as well as with nontechnical team members

Related Career Pathways/Majors

Information Technology Career Cluster
- Programming & Software Development Pathway

Science, Technology, Engineering & Mathematics Career Cluster
- Engineering & Technology Pathway

Postsecondary

Entry-level computer hardware engineers typically need a bachelor's degree in computer engineering or a related field, such as computer and information technology. Employers may prefer to hire candidates who have graduated from an engineering program accredited by a professional association, such as the Accreditation Board for Engineering and Technology, Inc. (ABET). To prepare for

Famous First

American engineer, entrepreneur, and medical doctor Henry Edward Roberts (1941-2010) invented the first commercially successful personal computer in 1974. Source: www.latimes.com

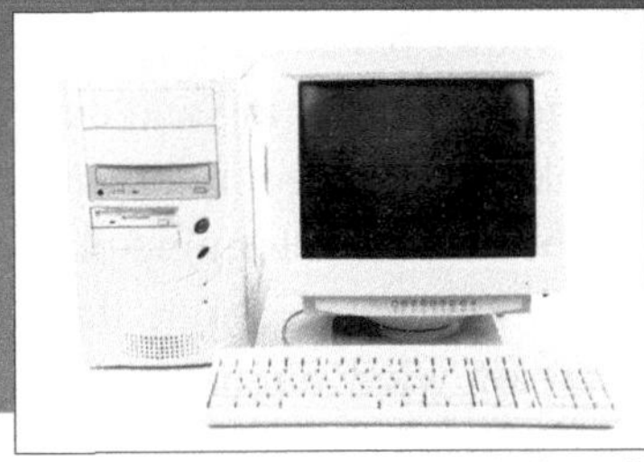

Photo via iStock/Devonyu. [Used under license.]

a major in computer or electrical engineering, students should have a solid background in math and science.

Because hardware engineers commonly work with computer software systems, a familiarity with computer programming is usually expected. This background may be obtained through computer science courses.

Some large firms or specialized jobs may require a master's degree in computer engineering. Some experienced engineers obtain a master's degree in business administration (MBA). All engineers must continue their learning over the course of their careers in order to keep up with rapid advances in technology.

Related College Majors

- Computer Engineering
- Computer Programming
- Computer Science
- Electrical Engineering
- Information Technology

Adult Job Seekers

If an adult job seeker with no prior experience in computer engineering or similar fields wishes to pursue such a career, they should enroll in a college that offers a program to expand their skills. As technology continues to advance, a computer hardware engineer must be open to continuously learning during the span of their career.

Professional Certification and Licensure

Although certification is not typically required by employers, there are organizations that offer certifications should an individual decide to pursue one.

Additional Requirements

Computer hardware engineers must exhibit exceptional analytical skills when it comes to examining and developing complex computer equipment. Critical thinking is an important skill for engineers in order for them to troubleshoot problems and determine proper solutions. Additionally, engineers must be willing to continue learning throughout their careers as technology advances and their knowledge must expand with it.

EARNINGS AND ADVANCEMENT

The median annual wage for computer hardware engineers was $119,560 in May 2020. The median wage is the wage at which half the workers in an occupation earned more than that amount and half earned less. The lowest 10 percent earned less than $70,100, and the highest 10 percent earned more than $192,110. Some computer hardware engineers can advance to become computer and information systems managers.

Depending on the employer, computer hardware engineers may receive sick days, paid vacations and holidays, as well retirement benefits, and health and life insurance.

EMPLOYMENT AND OUTLOOK

Computer hardware engineers held 66,200 jobs in 2020. Fifty-one percent were self-employed. Employment is expected to show slow change through the year 2030.

Demand for these engineers is expected to grow as more industries outside of the computer and electronic product manufacturing industry begin to research and develop their own electronic devices. However, centralized computing and networking services may mean fewer engineers will be needed because these worksites cover large networks and geographic areas.

An increase in hardware startup firms and the ongoing increase in the number of devices with computer chips embedded in them, such as household appliances, medical devices, and automobiles, may also lead to some job growth for these workers.

Related Occupations

- Aerospace Engineer
- Computer/Information Research Scientist
- Computer/Information Systems Manager
- Computer Network Architect
- Computer Programmer
- Electrical/Electronics Engineer
- Mechanical Engineer
- Network/Computer Systems Administrator
- Software Developer/Quality Assurance Analyst/Tester

MORE INFORMATION

Association for Computing Machinery (ACM)
1601 Broadway, 10th Floor
New York, NY 10019-7440
212.869.7440
www.acm.org

Association for Women in Computing (AWC)
P.O. Box 2768
Oakland, CA 94602
info@awc-hq.org
www.awc-hq.or

Computer Research Association (CRA)
1828 L Street
Washington, D.C. 20036-4632
202.234.2111
cra.org

IEEE Computer Society
2001 L Street, NW, 700
Washington, D.C. 20036-4928
202.371.0101
help@computer.org
www.computer.org

Conversation With...
NATALIE ENRIGHT JERGER

Professor
Edward S. Rogers Sr. Department of Electrical and Computer Engineering, University of Toronto
In the field, 20 years

What was your individual career path in terms of education/training, entry-level job, or other significant opportunity?

I completed my Bachelor's in computer engineering from Purdue University. While in undergrad, I interned twice at Hewlett Packard. After completing my degree, I enrolled in the PhD program at the University of Wisconsin-Madison. During my PhD, I interned at Intel and IBM. After completion of my PhD, I joined the University of Toronto as an assistant professor, where I have been since then.

What are the most important skills and/or qualities for someone in your profession?

In terms of technical skills, strong foundations in math and physics are very important. Programming is a pervasive skill in engineering at well. Teamwork and collaboration are also tremendously important skills. Strong communication skills are also important.

What do you wish you had known going into this profession?

Probably a lot. When I went into the profession, I knew relatively little about what computer engineering was, or what computer engineers did. I was just inherently curious to understand how computers worked. Knowing up-front that establishing a strong foundation in electrical and computer engineering, and establishing a breadth of knowledge, would have been useful. I think it's natural to focus on topics and skills that one enjoys, and at the start of one's career you can't always see how broadly useful various other skills might be.

Are there many job opportunities in your profession? In what specific areas?

There are many great job opportunities in computer engineering. Focusing on my area of computer hardware and computer architecture, there are lots of great jobs at traditional processor companies like Intel and AMD. In the last few years, there has been a rise of start-ups also building novel hardware to run key, emerging applications such as machine learning. Companies such as Microsoft, Facebook, and Google also require engineers with computer hardware expertise to design and manage their datacenters. A strong understanding of computer hardware can also make one a better programmer and software engineer so there are tons of career opportunities there as well.

How do you see your profession changing in the next five years, how will technology shift, and what skills will be required?

Moore's Law, which has been a major driver of innovation in computer hardware, has been slowing considerably in recent years, and will continue to slow down over the next 5 years. For decades, Moore's Law dictated that transistor sizes would halve every 1.5 to 2 years. As a result, we could double the number of transistors on each chip with every process generation. Transistors are the fundamental building blocks of digital systems and computer hardware. More transistors have resulted in more powerful computers which have, in turn, enabled more complex software. It has also allowed computers to become smaller and cheaper, which has resulted in more ubiquitous use of computers, such as through smart phones. The slowing of Moore's Law has opened the door to more specialized hardware such as chips detected to machine learning.

I think we will see more innovations in tailoring hardware to important applications. Computer hardware will continue to underpin our ability to conduct critical and important research across all disciplines, including in areas of healthcare, climate change, etc. Moving forward, computer engineers will require a broad skillset to understand all aspects of the system, from materials and circuits to hardware and software. Computer engineers should also be cognizant of the ethical implications of technology, as we consider the potential societal harms that can arise, such as bias, inequity, disinformation, and environmental harms.

What do you enjoy most about your job? What do you enjoy least about your job?

As an educator, it's definitely that "ah-ha" moment my students have, when they finally see how all the different topics they've been studying fit together to create a working computer. Some aspects of solving engineering problems often feel like puzzles—how do I fit the different pieces together, balance different objectives, etc.? I enjoy solving puzzles, so that aspect is also a lot of fun. In terms of what I enjoy least, it would be failure. As an engineer, design is an iterative process. Many ideas don't work on the first iteration and you have to get comfortable with failure, become good at debugging and identifying problems and iterating over your design.

Can you suggest a valuable "try this" for students considering a career in your profession?

I think there are a lot of great resources now for students to try out engineering projects at a young age. Building circuits or something with a simple micro-controller would be great. Try out some programming. For example, in one of the classes I have taught, students build cars, simple robots, etc., out of Lego and program them. Seeing your project move and interact is super exciting, and simple versions of this can be done prior to university.

Construction & Building Inspector

Snapshot

Career Cluster(s): Architecture & Construction; Manufacturing

Interests: Architecture, Physical Science, Engineering

Earnings (Yearly Average): $62,860

Employment & Outlook: Decline

OVERVIEW

Sphere of Work

Construction and building inspectors ensure that construction meets local and national building codes and ordinances, zoning regulations, and contract specifications. Construction and building inspectors ensure safety compliance of buildings, dams, bridges, and other structures; highways and streets; and sewer and water systems. They also inspect electrical; heating, ventilation, air-conditioning, and refrigeration (HVACR); and plumbing systems. Inspectors typically check a project several times: for an initial check in the early construction phase, for follow-up inspections as the project progresses, and for a comprehensive examination after its completion. At each inspection, they may provide written or oral feedback about their findings.

Photo via iStock/microgen. [Used under license.]

Work Environment

Although construction and building inspectors spend most of their time examining worksites, they also spend time in an office reviewing blueprints, writing reports, and scheduling inspections. Some inspectors climb ladders or crawl in tight spaces as part of their work. Inspectors typically work alone. However, inspectors may work as part of a team on large, complex projects, particularly if they specialize in one area of construction.

Profile

Working Conditions: Both Inside and Outside
Physical Strength: Light Work
Education Needs: High School Degree; On-the-Job Training; Apprenticeship
Licensure/Certification: Required, Varies Between States
Opportunities for Experience: Internship, Apprenticeship, Part-Time Work
Interest Score: RCI

Occupation Interest

The field of building inspection often attracts those with backgrounds in engineering, physical science, architecture, and civic planning. Many inspectors arrive at the position after several years in the private construction industry, either as skilled laborers or as engineering consultants, project managers, or architects.

Construction is a multifaceted discipline that requires knowledge of an array of logistics and systematic infrastructure, including architecture, HVACR, plumbing, electrical circuitry, weatherproofing, load-bearing metrics, and aerodynamics. Inspectors must also be well-versed in local, state, and national building regulations.

Duties and Responsibilities

- Reviewing building plans and approving those that meet requirements
- Monitoring construction sites periodically to ensure overall compliance
- Using equipment and testing devices, such as moisture meters to check for plumbing leaks or flooding damage and electrical testers to ensure that electrical components are functional
- Inspecting plumbing, electrical, and other systems to ensure that they meet code
- Issuing violation notices and stop-work orders if building is not compliant
- Keeping daily logs, which may include digital images from inspectors
- Documenting findings in writing

A Day in the Life—Duties and Responsibilities

Much of the day-to-day responsibilities of building inspectors involve traveling to and inspecting construction sites. The scale, location, and breadth of site surveys will depend on the specialty of the inspector.

Civic building inspectors survey all new construction sites and renovation projects within their particular jurisdictions to ensure that the projects fall within the parameters of regional, state, and federal building codes.

Construction inspectors spend much of their noninspection time educating themselves about alterations to existing codes as well as new building codes, which are traditionally issued on an annual basis. In many cases, inspectors are required to attend conferences and seminars where new building codes or code-friendly building techniques are taught.

Civic inspectors possess the capacity to halt construction projects that are in violation of building codes. Reasons for a building inspector to shut down construction can range from the use of illegal materials to improper site waste management, hazardous work conditions, improper implementation of safety equipment, or repetition of a combination of such offenses. Construction projects that are halted by inspectors are required to reapply for building certificates and often must pass thorough inspections prior to being allowed to proceed. Building inspectors are often called upon to interpret construction laws and building codes for project managers eager to preempt a disruption of progress.

OCCUPATION SPECIALTIES

Building Inspector
Building inspectors check the structural quality, architectural requirements, and general safety of buildings. Some building inspectors focus on fire prevention and safety.

Coating Inspector
Coating inspectors examine the exterior paint and coating on bridges, pipelines, and large holding tanks. In their checks throughout the painting process, inspectors ensure that protective layers are correctly applied.

Electrical Inspector
Electrical inspectors examine a building's installed electrical systems to ensure compliance and proper functioning. These systems may include new and existing sound and security systems, lighting, photovoltaic systems, generating equipment, and wiring for HVACR systems and appliances.

Elevator Inspector
Elevator inspectors examine lifting and conveying devices, such as elevators, escalators, moving sidewalks, lifts and hoists, inclined railways, ski lifts, and amusement rides. They inspect both the mechanical and electrical control systems.

Home Inspector
Home inspectors typically examine houses, condominiums, townhomes, and other dwellings to report on their structure and overall condition. Home sellers or home buyers, or both, may seek inspectors' objective assessment of a dwelling before placing it on the market or submitting an offer.

Mechanical Inspector
Mechanical inspectors examine heating, ventilation, and air conditioning (HVAC) systems and equipment to ensure that they are installed and function properly. They also may inspect commercial kitchen equipment, gas-fired appliances, and boilers. Mechanical inspectors' work differs from that of quality control inspectors, who inspect goods at manufacturing plants.

Plans Examiner
Plans examiners determine whether the plans for a building or other structure comply with adopted building codes, regulations, and ordinances.

Plumbing Inspector
Plumbing inspectors examine the installation of systems that ensure the safety of drinking water and industrial piping and the sanitary disposal of waste.

Public Works Inspector
Public works inspectors ensure that the construction of federal, state, and local government water and sewer systems; roads and bridges; and dams conforms to specifications. They may specialize in projects such as highways, structural steel, or dredging operations required for bridges, dams, or harbors.

Special Inspector
Special inspectors ensure that critical construction work, such as high-strength concrete, steel fabrication, and welding, is installed and tested according to design specifications. Special inspectors represent the owner's interests, not those of the general public. Insurance companies and financial institutions also may use their services.

WORK ENVIRONMENT

Immediate Physical Environment

Construction and building inspectors predominantly work on construction sites. They also often work outdoors to check the exterior structures of whichever project they are working on, which can range from large-scale civic engineering projects like highways or bridges, to smaller-scale projects like residential houses or worksites. Furthermore, construction and building inspectors must understand and be comfortable that they may be working in potentially dangerous or hazardous construction sites, or difficult climates or environments.

Human Environment

Typically, inspectors tend to work alone in their projects. However, they may also work alongside colleagues and work as part of a team when it comes to large and complex projects. Especially if an inspector specializes in one area of construction and their expertise is needed in a larger project.

Technological Environment

Construction inspectors must be well-versed in a variety of measurement systems and analytical tools measuring corrosion, exposure, and temperature. Advanced mathematic skills are also highly beneficial.

EDUCATION, TRAINING, AND ADVANCEMENT

High School/Secondary

The best way for those interested in a career path of a construction and building inspector is by participating in courses such as algebra, chemistry, geometry, physics, and computer science. If a school has courses or extracurricular related to architecture then it would be beneficial for those interested in entering the field to partake in them. Additionally, gaining work experience through a summer job or school internships can be another way to get one's feet in the door and look good on a resume.

Suggested High School Subjects

- Algebra
- Calculus
- Chemistry
- Construction Math
- English
- Environmental Science
- Geometry
- Physics
- Trigonometry
- Woodshop

Related Career Pathways/Majors

Architecture & Construction Career Cluster

- Construction Pathway
- Maintenance & Operations Pathway
- Quality Assurance Pathway

Postsecondary

A postsecondary degree will vary between employers where some would prefer an individual with a degree while another may not. Those who choose to pursue a secondary education often come into the profession in similar fields such as architecture, civil planning, and engineering. A degree may prove beneficial to acquire for someone interested in becoming a building inspector because during this time they will study the fundamentals of construction and engineering, as well as potentially gaining experience through internships with private building, engineering, or construction firms.

Some employers may seek candidates who have a bachelor's degree in engineering or architecture or who have another postsecondary credential. Many community colleges offer a certificate or an associate's degree program in building inspection technology and have courses in building inspection, home inspection, construction technology, and drafting. Courses in blueprint reading, vocational subjects, algebra, geometry, and writing are also useful. Courses in business management are helpful for those who plan to run their own inspection business.

Transferable Skills and Abilities

Communication Skills

- Explaining issues or problems they discover and writing a report that clearly describes their findings
- Thoroughly examining different construction activities
- Paying close attention to the activities in order to not overlook any details

Mechanical Knowledge

- Using a variety of testing equipment to check complex systems
- Understanding how the systems operate

Physical Stamina

- Frequently on their feet and climbing or crawling through attics and other tight spaces

Related College Majors

- Architecture
- Civil Engineering
- Construction Technology
- Drafting and Design Technology
- Engineering Technology
- Surveying

Adult Job Seekers

Adult job seekers can prepare for a career in building inspection by gaining sustained experience in the engineering field, particularly at the supervisory level. Leadership and managerial experience in construction management is also desirable. Many inspectors have several years of experience working as carpenters, systems engineers, or electricians. Sound collaborative and communication skills are also paramount for those seeking careers in the field. Candidates with multidisciplinary backgrounds in construction often have broadened opportunities.

Professional Certification and Licensure

Many states and localities require some type of license or certification. Additionally, some jurisdictions require that construction and building inspectors take continuing education course to maintain their credentials as building codes and legal requirements constantly change.

Additional Requirements

Training requirements vary by state, locality, and type of inspector. In general, construction and building inspectors receive much of their training on the job. Construction and building inspectors learn building codes and standards as a prerequisite to obtaining their license and through continuing education. Working with an experienced inspector, they learn about inspection techniques; codes, ordinances, and regulations; contract specifications; and recordkeeping and reporting duties. Training also may include supervised on-site inspections.

EARNINGS AND ADVANCEMENT

The median annual wage for construction and building inspectors was $62,860 in May 2020. The median wage is the wage at which half the workers in an occupation earned more than that amount and half earned less. The lowest 10 percent earned less than $37,850, and the highest 10 percent earned more than $101,170.

Most inspectors work full time during regular business hours. However, some work additional hours during periods of heavy construction. Also, if an accident occurs at a construction site, inspectors must respond immediately and may work additional hours to complete their report. Some inspectors—especially those who are self-employed—work evenings and weekends. This is particularly true of home inspectors, who typically inspect homes during the day and write reports in the evening.

EMPLOYMENT AND OUTLOOK

Employment of construction and building inspectors is projected to decline 3 percent from 2020 to 2030. Despite declining employment, about 14,300 openings for construction and building inspectors are projected each year, on average, over the decade. All of those openings are expected to result from the need to replace workers who transfer to other occupations or exit the labor force, such as to retire.

The increasing use of remote inspections will reduce the need for inspectors in state and local government. However, continued public interest in safety and the desire to improve the quality of construction are among the factors expected to create demand for inspectors.

Related Occupations

- Architect
- Carpenter
- Construction Manager
- Electrical/Electronic Engineering Technologist/Technician
- Electrical/Electronics Engineer
- Electrician
- Occupational Health/Safety Specialist/Technician
- Plumber, Pipefitter, & Steamfitter
- Property Appraiser/Assessor
- Surveyor

MORE INFORMATION

American Society of Home Inspectors (ASHI)
932 Lee Street, Suite 101
Des Plaines, Illinois 60016-6594
847.759.2820
www.homeinspector.org

Association of Construction Inspectors (ACI)
P.O. Box 879
Palm Springs, CA 92263
877.743.6806
www.aci-assoc.org

International Association of Certified Home Inspectors (InterNACHI)
1750 30th Street, Suite 301
Boulder, CO 80301
888.422.7233
fastreply@internachi.org
internachi.org

International Association of Electrical Inspectors (IAEI)
901 Waterfall Way, Suite 602
Richardson, TX 75080
800.786.4234
www.iaei.org

International Association of Plumbing and Mechanical Officials (IAPMO)
4755 E. Philadelphia St.
Ontario, CA 91761
909.472.4100
www.iapmo.org

International City/County Management (ICMA)
777 N. Capitol Street, NE, Suite 500
Washington D.C. 20002-4201
800.745.8780
icma.org

International Code Council (ICC)
500 New Jersey Avenue, NW, 6th Floor
Washington, DC 20001
888.422.7233
www.iccsafe.org

National Association of Elevator Safety Authorities International (NAESAI)
P.O. Box 4367
Mankato, MN
360.292.4968
naesai.org

National Fire Protection Association (NFPA)
1 Batterymarch Park
Quincy, MA 02169-7471
800.344.3555
www.nfpa.org

Kristina Domizio & John Pritchard

Conversation With...
JOE HALL

Owner
Joe Hall Energy Consulting, LLC, Denver, CO
Building Inspections, 25 years

What was your individual career path in terms of education/training, entry-level job, or other significant opportunity?

Back in 1993, I just needed a job. I started as an entry-level weatherization technician, air-sealing and insulating homes for a non-profit. I never dreamed it would turn into a career and take me all over the country.

About a year after I started, I became an energy auditor, then a field supervisor, supervising energy auditors and weatherization technicians. For 10 years after that, I installed furnaces as an HVAC technician working for Longs Peak Energy Conservation in Boulder County. Then I went to a now-defunct nonprofit called Sunpower and ran the state of Colorado's training center. Under the American Recovery and Reinvestment Act of 2009, my industry got $400 billion over a two-year period.

I was working for people who weren't doing much work, but they were making thousands and I was making hundreds. So I cut out the middle man and started my own company. We do inspections as well as training and consulting. People call and say, "My house isn't working." My job is finding solutions. We first do a health and safety inspection. We need to identify any problems before we do weatherization, because if that's done wrong, we can exacerbate problems. I work with realtors with $5 million homes and plain old regular people with 1,200 square foot homes. And I work with the low-income sector, working for government.

On the training side, I do classes on building science—going beyond the electrical box and shingles—for the International Association of Certified Home Inspectors (InterNACHI); I train workers for utility rebate programs and do trades programs in prisons: I've piloted programs for the National Renewable Energy Laboratory and worked with the government on the Healthy Homes Initiative and the Home Energy Score. The short version of what I do is teach physics to building trades folks. My own training began with a weeklong class that the state offered, then I did more training with On the Job Training (OJT). Throughout my career, I've done a lot more training, gotten a lot of certifications, attended conferences, and taken online courses with the Building Performance Institute (BPI).

What are the most important skills and/or qualities for someone in your profession?
A genuine understanding of all aspects of building trades. You have to understand the mechanicals, you have to understand framing, you have to understand ventilation and you have to understand how all those subsystems interact.

What do you wish you had known going into this profession?
I wish I had found it sooner. I did a lot of different things before this. I find this interesting. Every house is different. I'm concerned about energy and energy conservation. Most of my customers are not; they're concerned about comfort. But if I make their house comfortable, it's going to be energy-efficient.

Are there many job opportunities in your profession? In what specific areas?
If somebody wanted to get a job installing insulation, I could get them a job tomorrow. That's a good place to start, in air-sealing and insulation then work your way up as I did. There are jobs for energy auditors and a lot of jobs doing training.

How do you see your profession changing in the next five years? What role will technology play in those changes, and what skills will be required?
Building code requirements necessitate that we use more technology and equipment, and therefore need expertise on how that equipment runs.

The rating system used by the U.S. Green Building Council to measure a building's sustainability and efficiency is known as LEED, or Leadership in Energy and Environmental Design. LEED certification has all the trades together, so, for instance, the HVAC people look beyond their box and start looking at the big box and all the other boxes that are in that big box and how they interact.

Also, the health care profession is starting to understand that building scientists can have a pretty good idea why people are getting sick. If someone keeps showing up with asthma, health professionals are realizing they can get someone like me to look over the home and write up a report so they can see if the environment is causing the illness.

What do you enjoy most about your job? What do you enjoy least about your job?
I like finding and solving people's house problems. I like educating consumers and training professionals. I like watching the lightbulb go on when people start to understand what I'm explaining to them.

What I like least is crawling underneath houses with sewer leaks. And I really don't like attics at 150 degrees. Also, sometimes educating people can be frustrating because building science is a complex subject—but it's still fun when you see you're making progress.

Can you suggest a valuable "try this" for students considering a career in your profession?
Go to The Home Depot and talk to people—the ones with the pallets, not the baskets. I've taught at Mile High Youth Corps, which helps non-profits and low-income people with energy efficiency. Do that or a similar AmeriCorps program. Check out the InterNACHI website, www.nachi.org. You could take online classes at www.bpi.org.

This interview was originally published in 2018.

Conversation With...
AUDREY CLINE

Code Official
Town of Stratham Building/Codes Department
Stratham, New Hampshire
Building inspector, 10 years

What was your individual career path in terms of education/training, entry-level job, or other significant opportunity?

Like many (if not most) building inspectors, code enforcement officers, and code officials, I landed in this field as a mid-career change. I graduated from Boston University with a degree in business administration. A handful of years later, I developed an interest in architecture and attended the architectural engineering program at New Hampshire Technical Institute. After graduation, I worked as a self-employed residential designer. After a decade or so, the work began to feel mundane and I starting casting about for another situation where I could use my skills and training.

I "discovered" the field of building codes when I saw an ad in my local paper for an employment opportunity with the Town of Wolfeboro, N.H. As the town's building official, I managed the building department's processes, which included reviewing plans for compliance with the New Hampshire State Building Code and local zoning ordinances and collaborating with local fire officials for compliance with the New Hampshire State Fire Code. I enjoy the legal aspects of code administration and property rights. My favorite projects are redevelopment or so-called "infill" projects that rejuvenate older buildings and neighborhoods while bringing them up to code to meet structural strength and safety regulations.

What are the most important skills and/or qualities for someone in your profession?

It's important to realize that the decisions you make may have a significant effect on personal property rights. It's critical to develop an approach based on the desire to be part of the solution, while fulfilling your responsibility to the public health, safety, and welfare. This is not a career for the faint-of-heart. In the end, credibility is the goal. With credibility, even difficult decisions can be understood as being necessary and fair.

What do you wish you had known going into this profession?

There isn't anything I regret about making this career move. What I didn't know, and what was most surprising to me, is that public service is very different from working in the private sector. I had to re-evaluate my basic decisionmaking and judgment processes in order to accommodate the notion of the public good rather than the good of an individual, or indeed, my own preference.

Are there many job opportunities in your profession? In what specific areas?
In New Hampshire as well as many other states, the next five to ten years will see a major shortage of people entering this field. In response to the expected shortage, New Hampshire Technical Institute has begun a certificate program designed to attract people to the industry. The program can lead to immediate employment after six courses or can be the foundation for stepping into an associate's or bachelor's degree program in a related field like construction management or architecture/engineering.

While building inspectors and code officials typically work for a city or town, there are home inspection jobs in private industry. When people buy a home, the bank typically requires a home inspection that covers the condition of the home from the foundation to the roof: heating and a/c systems, plumbing and electrical systems, attic/insulation, ceiling, windows and floors. Home inspectors inspect these systems and write up a report for the buyer.

How do you see your profession changing in the next five years? What role will technology play in those changes, and what skills will be required?
Just as in fields like medicine and technology, there will be distinct sub-specialties. There will be a need for management personnel who have a thorough grasp of technical codes, as well as the education (such as a degree in public administration) to manage a municipal office. Most building offices are instituting electronic applications, plan review, and inspection protocols.

What do you enjoy most about your job? What do you enjoy least about your job?
As the town's building official, I am just as apt to be outside on a job site as in the plan review room or attending a policy committee meeting. I love the variety. Being a resource for builders and potentially saving a project from an expensive or time-consuming retrofit is very satisfying. It can be discouraging to see a project that has derailed because of lack of proper design, planning or implementation. But after ten years, I am still waiting to learn what I least like about my job!

Can you suggest a valuable "try this" for students considering a career in your profession?
I would urge interested students to contact the association of building officials in their state. If you're in New Hampshire or nearby, contact the New Hampshire Building Officials Association (www.nhboa.net). The association has the ability to set up a shadowing program and will reserve a seat for a curious student at one of our monthly training meetings. I'm sure other state organizations have similar programs.

This interview was originally published in 2016.

Drafter

Snapshot

Career Cluster(s): Architecture & Construction; Science, Technology, Engineering & Mathematics

Interests: Architecture; Mathematics; Drawing; Engineering

Earnings (Yearly Average): $57,960

Employment & Outlook: Slower than Average

OVERVIEW

Sphere of Work

Drafters use software to convert the designs of architects and engineers into technical drawings. Most workers specialize in architectural, civil, electrical, or mechanical drafting and use technical drawings to help design everything from microchips to skyscrapers.

Work Environment

Drafters spend much of their time working on computers using specialized software inside a professional engineering or architectural office. Their computers are usually located at their designated workspace that is often a large desk that allows for sufficient space to lay out design work for examination and alterations. Alongside their time spent at their work desk, drafters may also visit jobsites to collaborate with architects and engineers.

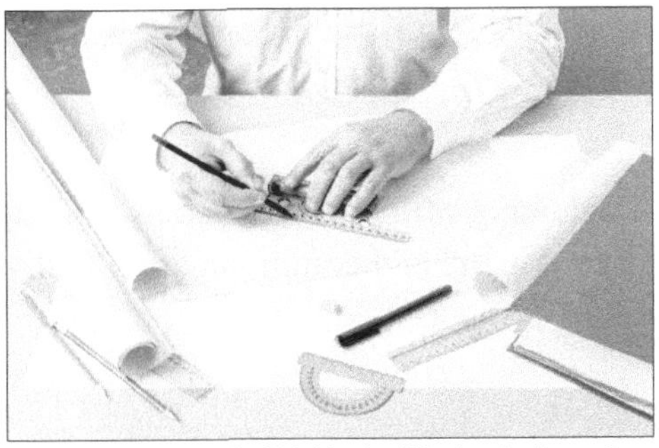

Photo via iStock/Viktoriia Oleinichenko. [Used under license.]

Occupation Interest

Individuals interested in a career in drafting often are very detail-oriented with a strong understanding of mathematics and engineering. Many engineering and architectural professionals start off as a drafter, and the role offers advancement into such fields.

Additionally, it would benefit and individual to enjoy drawing with an ability to judge perspective and dimensions when it comes to technical drawings. Time management skills are also necessarily to possess when it comes to deadline sensitive projects that drafters are often faced with.

Profile

Working Conditions: Mainly Indoors
Physical Strength: Light Work
Education Needs: Associate's Degree; Bachelor's Degree; Certification
Licensure/Certification: Not Required but Preferred
Opportunities for Experience: Internship; Apprenticeship
Interest Score: RCA

A Day in the Life—Duties and Responsibilities

Drafters typically work as part of a design-and-development team that translates the ideas of engineers into designs for construction or manufacturing. They generally work during regular business hours, and the set of tasks for any given day may vary considerably.

Duties and Responsibilities

- Designing plans using computer-aided design and drafting (CADD) software
- Working from rough sketches and specifications created by engineers and architects
- Designing products with engineering and manufacturing techniques
- Adding details to architectural plans from knowledge of building techniques
- Specifying dimensions, materials, and procedures for new products
- Working under the supervision of engineers or architects

Typically, drafters consult with project managers, engineers, and architects during a typical business day to plan and schedule projects.

Drafters may spend part of a business day translating paper drawings to computer-aided design and drafting (CADD) format, or they may work directly in CADD to design a new schematic. A certain amount of time must also be spent updating or refining existing CADD designs following a review and incorporating additional requirements determined by the engineers. Some time might be spent conducting office work, such as sending schematics or design templates to clients or customers and communicating with others in the office about ongoing and future projects.

OCCUPATION SPECIALTIES

Architectural Drafter

Architectural drafters draw structural features and details for buildings and other construction projects. These workers may specialize in a type of building, such as residential or commercial. They may also specialize by the materials used, such as steel, wood, or reinforced concrete.

Civil Drafter

Civil drafters prepare topographical maps used in construction and civil engineering projects, such as highways, bridges, and dams.

Electrical Drafter

Electrical drafters prepare wiring diagrams that construction workers use to install and repair electrical equipment and wiring in power plants, electrical distribution systems, and residential and commercial buildings.

Electronics Drafter

Electronics drafters produce wiring diagrams, assembly diagrams for circuit boards, and layout drawings used in manufacturing and in installing and repairing electronic devices and components.

Mechanical Drafter

Mechanical drafters prepare layouts that show the details for a variety of machinery and mechanical tools and devices, such as medical equipment. These layouts indicate dimensions, fastening methods, and other requirements for assembly. Mechanical drafters sometimes create production molds.

WORK ENVIRONMENT

Immediate Physical Environment

The work environment for a drafter is usually an indoor office that is bright, open, and well-ventilated. Drafters would often be supervised by engineers or architects whose projects they are adapting.

Human Environment

Depending on the size of the office, an individual could be the only drafter present on a team, or if it's a larger architectural or engineering firm then there may be multiple drafters to work between numerous projects. Additionally, drafters are adapting designs into technical drawings, so they would be working alongside the engineers or architects behind the design. Communication between other members of the team is to be expected.

Technological Environment

Professional drafters utilize some form of CADD software and may use a variety of other programs depending on the nature of their work. They occasionally use three-dimensional (3D) modeling software to create designs for various products or component parts. There are also project-management software programs that allow drafters, engineers, and other specialists to collaborate on designs through a computer network.

EDUCATION, TRAINING, AND ADVANCEMENT

High School/Secondary

High school students may begin preparing by taking classes in mathematics, science, computer technology, design, computer graphics, and, where available, drafting. Additionally, students wishing to pursue a career in crafting can gain and sharpen their skills through technical drawing and art classes. Students can also enroll in their school's computer science program to better build up knowledge on the different types of software that are used in drafting.

Suggested High School Subjects

- Algebra
- Calculus
- Computer Graphics
- Computer Technology
- Design
- Drafting
- English
- Geometry
- Mechanical Drawing
- Trigonometry

Related Career Pathways/Majors

Architecture and Construction Career Cluster

- Design/Pre-Construction Pathway

Science, Technology, Engineering and Mathematics Career Cluster

- Engineering & Technology Pathway

Postsecondary

Drafters typically complete education after high school, often through a program at a community college or technical school. Some programs lead to an associate of applied science in drafting or a related degree; others result in a certificate or diploma.

Drafters typically need an associate of applied science in drafting or a related degree from a community college or technical school. Some drafters prepare for the occupation by earning a certificate or diploma. Programs in drafting may include instruction in design fundamentals, sketching, and CADD software. It generally takes about two years of full-time education to earn an associate's degree. Certificate and diploma programs vary in length but usually may be completed in less time. Students frequently specialize in a particular type of drafting, such as mechanical or architectural drafting.

Transferable Skills and Abilities

Creativity

- Turning plans and ideas into technical drawings of buildings, tools, and systems

Detail-oriented

- Taking care that the plans they are converting are technically accurate according to the outlined specifications

Interpersonal Skills

- Communicating effectively with fellow architects, engineers, and other designers to make sure that final plans are accurate

Math Skills

- Accurately calculating angles, weights, costs, and other values on their technical drawings

Technical Skills

- Understanding the computer software they use, such as CAD
- Working with database tools, such as building information modeling (BIM)

Time-management Skills

- Working effectively under strict deadlines to produce the required output according to set schedules

Related College Majors

- Architecture
- Civil Drafting
- Civil Engineering
- Drafting & Design
- Mechanical Engineering

Adult Job Seekers

Adults interested in pursuing a career as a drafter but lack the experience can pursue classes at a nearby community college or technical school that offers drafting. If an adult job seeker isn't familiar with computers or design programs that will be used as a drafter, then they may find it in their interest to take a class on basic computer skills that start from the basics to digital design programs.

Professional Certification and Licensure

The American Design Drafting Association (ADDA) offers certification for drafters. Although not mandatory, certification demonstrates competence and knowledge of nationally recognized practices. Certifications are offered for several specialties, including architectural, civil, and mechanical drafting.

Additional Requirements

Drafters must be creative individuals who are able to turn designs and plans given to them by engineers and architects into technical drawings. Furthermore, they have to be highly motivated and self-starters with the ability to work independently on projects while communicating clearly with their supervisors, as well as fellow architects and engineers, or drafters they work with.

EARNINGS AND ADVANCEMENT

The median annual wage for drafters was $57,960 in May 2020. The median wage is the wage at which half the workers in an occupation earned more than that amount and half earned less. The lowest 10 percent earned less than $36,470, and the highest 10 percent earned more than $87,970.

EMPLOYMENT AND OUTLOOK

Drafters held 191,800 jobs in 2020. Overall employment of drafters is projected to decline 2 percent from 2020 to 2030. Despite declining employment, about 17,500 openings for drafters are projected each year, on average, over the decade. All of those openings are expected to result from the need to replace workers who transfer to other occupations or exit the labor force, such as to retire.

Expected employment decreases in manufacturing and engineering services will more than offset the small increases in construction. These decreases will be driven by the use of CADD and BIM technologies, which allow engineers and architects to perform many tasks that used to be done by drafters.

Related Occupations

- Architect
- Civil Engineering Technologist and Technician
- Electrical/Electronic Engineering Technologist/Technician
- Electrical/Electronics Engineer

Fast Fact

The 1960s saw a boom in technologies that simplified technical drawing, especially those with complex designs—known as computer-aided design (CAD)—which continue to change the way drafters and other design professionals work. Building Information Modeling (BIM) is also revolutionizing design, and involves creating an intelligent 3D model that guides the lifespan of a project.

Source: engineerwarehouse.com

Photo via iStock/gorodenkoff. [Used under license.]

- Industrial Designer
- Landscape Architect
- Mechanical Engineer
- Mechanical Engineering Technologist/Technician
- Surveying and Mapping Technician
- Surveyor

MORE INFORMATION

American Council of Engineering Companies (ACEC)
1015 15th Street, NW, 8th Floor
Washington, DC 20005-2605
202.347.7474
acec@acec.org
www.acec.org

American Design Drafting Association (ADDA)
105 E. Main Street
Newbern, TN 38059
731.627.0802
www.adda.org

American Institute of Architects (AIA)
1735 New York Avenue, NW
Washington D.C. 20006-5292
www.aia.org

Association of Collegiate Schools of Architecture (ACSA)
1735 New York Avenue
Washington, DC 20006
202.785.2324
info@acsa-arch.org
www.acsa-arch.org

Society of American Registered Architects (SARA)
14 E. 38th Street, 11th Floor
New York, NY 10016
212.520.2901
www.sara-national.org

Kristina Domizio & Micah L. Issitt

Conversation With...
KIM PRITCHARD

Mechanical Drafter/Designer
Bard Medical, Salt Lake City, Utah
Drafting and design, 10 years

What was your individual career path in terms of education/training, entry-level job, or other significant opportunity?

I got my Associate of Applied Science in engineering graphics and design technology from Utah Valley University. My first career-related job was as a part-time drafter using SolidWorks software while I was still in school. From there, I bounced through a few drafting jobs until I got into the medical device industry as a mechanical drafter/designer. I've been in that field ever since and am now finishing up my bachelor's in technology management. If you can, get a bachelor's degree. I didn't have to pay for mine because my employers have all had tuition reimbursement policies.

What are the most important skills and/or qualities for someone in your profession?

You need to be a visual person. You need to be organized, clean, and precise. Your drawings must leave nothing to the imagination; they must have every piece of needed information so that those who use them to build, write quotes, etc., won't have to call you about some important detail that's missing. You need to be able to work with others, particularly engineers. They will ask for something and often, you'll need to press them for more information. They sometimes provide vague details expecting you to fill in the blanks. You won't be able to fill in the blanks on your own; you'll need them to help you.

It helps to be open-minded and willing to learn new things, especially when it comes to design. As you get more experience, you'll learn how to design as engineers do. That is where your value increases, so don't be afraid to push yourself into taking on new roles. Some drafters are just that, drafters. They come to work, take orders, churn out simple drawings, and go home. Learn to grow your skills, learn new CAD programs, and learn engineering concepts. Most importantly, evolve into a designer, which will earn you more respect and money. You can be making as much as an engineer, with half the schooling!

What do you wish you had known going into this profession?

You will not be recognized as a full-blown engineer. Engineers go to school for four to eight years and earn the recognition and responsibility of that title. In school, I thought that drafters evolved into engineers with experience—this is false. You may gain experience over time that will make it seem like you're an engineer, but without that bachelor's degree in engineering, you're not one.

A big goal for me was making lots of money, so I pressed people on which drafting fields made the most money. Architectural? Mechanical? Structural? From what I've seen, electrical drafters make the most money. Electrical is ingrained in high technology and construction, which are both lucrative industries. A friend who has the same degree as I do makes over $100,000 a year doing electrical building information modeling (BIM) work. Structural steel drafters also do well.

Are there many job opportunities in your profession? In what specific areas?
As long as the world needs drawings, there will be a need for drafters. There are opportunities everywhere, in any field you can think of. A basic 2D drafter can find a job almost anywhere for about $18 an hour. If you want to make more, you need to specialize and become a designer. Then your value goes up, because you're no longer just putting basic lines on paper; you are designing components, systems, layouts, and the like.

How do you see your profession changing in the next five years? What role will technology play in those changes, and what skills will be required?
The world is moving into 3D and beyond; 2D is dead. Learn to think in 3D. Learn parametric software programs like SolidWorks, Creo, Inventor, and Revit. AutoCAD, which is 2D, is still around, but it's becoming a relic. Rarely will you find a high-paying, complex job that uses AutoCAD.

What do you enjoy most about your job? What do you enjoy least about your job?
What I enjoy most is creating and drawing. You don't have the responsibilities of engineers, but you can make as much money. In my experience, drafters are rarely on the hot seat. People tend to leave you alone. The engineers have to deal with all the difficult stuff. The downside is drafters often can be overlooked. When it comes time for accolades and awards, prepare to never see your name come up. You can often feel undervalued and under-appreciated.

Can you suggest a "try this" for students considering a career in your profession?
If you like to draw and create, drafting and design is for you. If you decide to pursue this field, get a drafting degree; drafting certificates are worthless. In your last year of school, get a part-time job as a drafter. I made $16 an hour as a part-time drafter in college, which, at the time, was excellent money!

This interview was originally published in 2016.

Electrical/Electronics Engineer

Snapshot

Career Cluster(s): Arts, A/V Technology & Communications; Science, Technology, Engineering & Mathematics

Interests: Electronics; Electrical Systems; Technology; Design; Creativity

Earnings (Yearly Average): $103,390

Employment & Outlook: As Fast As Average Growth Expected

OVERVIEW

Sphere of Work

Electrical engineers design, develop, test, and supervise the manufacture of electrical equipment, such as electric motors, radar and navigation systems, communications systems, or power generation equipment. Electrical engineers also design the electrical systems of automobiles and aircraft.

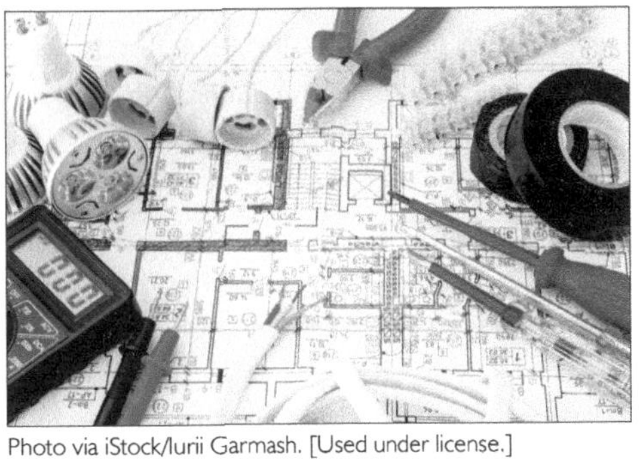

Photo via iStock/Iurii Garmash. [Used under license.]

Electronics engineers design and develop electronic equipment, including broadcast and communications systems, such as portable music players and Global Positioning System (GPS) devices. Many also work in areas closely related to computer hardware.

Work Environment

Electrical and electronics engineers typically work in offices of various sizes, although travel may be necessary to visit sites where their equipment is being installed and operated.

Profile

Working Conditions: Both Inside and Outside
Physical Strength: Medium Work
Education Needs: Bachelor's Degree
Licensure/Certification: Usually Not Required
Opportunities for Experience: Internship; Co-op
Interest Score: IR

Occupation Interest

Electrical and electronics engineers are creative, math- and science-oriented individuals who want to apply their knowledge to the creation of new electrical and electronic systems, components, and equipment. Applications in this field are wide-ranging, so engineers are often able to apply their skillset to fields they are passionate about.

A Day in the Life—Duties and Responsibilities

The work of electrical engineers and electronics engineers is often similar. Both use engineering and design software and equipment to do engineering tasks. Both types of engineers also must work with other engineers to discuss existing products and possibilities for engineering projects.

Electronics engineers who work for the federal government research, develop, and evaluate electronic devices used in a variety of areas, such as aviation, computing, transportation, and manufacturing. They work on federal electronic devices and systems, including satellites, flight systems, radar and sonar systems, and communications systems.

Duties and Responsibilities

Electrical Engineer

- Designing new ways to use electrical power to develop or improve products
- Performing detailed calculations to develop manufacturing, construction, and installation standards and specifications
- Directing the manufacture, installation, and testing of electrical equipment to ensure that products meet specifications and codes
- Investigating complaints from customers or the public, evaluating problems, and recommending solutions
- Working with project managers on production efforts to ensure that projects are completed satisfactorily, on time, and within budget

WORK ENVIRONMENT

Immediate Physical Environment

Electrical and electronics engineers generally work indoors in offices. However, they may visit sites to observe a problem or a piece of complex equipment. Most electrical and electronics engineers work full-time.

Duties and Responsibilities

Electronics Engineer

- Designing electronic components, software, products, or systems for commercial, industrial, medical, military, or scientific applications
- Analyzing customer needs and determining the requirements, capacity, and cost for developing an electrical system plan
- Developing maintenance and testing procedures for electronic components and equipment
- Evaluating systems and recommending design modifications or equipment repair
- Inspecting electronic equipment, instruments, and systems to make sure they meet safety standards and applicable regulations
- Planning and developing applications and modifications for electronic properties used in parts and systems to improve technical performance

Human Environment

Electrical and electronics engineers may conduct solitary office and laboratory work, but team efforts are common, and engineers must be prepared to work with manufacturers, visit sites, interact with customers and the public, and troubleshoot issues, sometimes under stressful conditions if their equipment is being used in time-sensitive circumstances.

Technological Environment

By their nature, electrical and electronics engineers must be intimately familiar with the equipment necessary to carry out their work and must stay apprised of new developments in their field. Computer-aided design (CAD) software is commonly used, as is analytical or scientific software, development environment software, and object- or component-oriented development software.

EDUCATION, TRAINING, AND ADVANCEMENT

High School/Secondary

High school students interested in studying electrical or electronics engineering benefit from taking courses in physics and math, including algebra, trigonometry, and calculus. Courses in drafting are also helpful, because electrical and electronics engineers often are required to prepare technical drawings. During high school, students can attend engineering summer camps to see what these and other engineers do. Attending these camps can help students plan their coursework for the remainder of their time in high school. The Engineering Education Service Center (EESC) has a directory of engineering summer camps.

Suggested High School Subjects

- Algebra
- Biology
- Chemistry
- Civics

- Computer Science
- Drafting
- Earth or Environmental Science
- Economics
- English
- Entrepreneurship
- Geometry
- History
- Physics
- Pre-Calculus
- Psychology
- Statistics
- Trigonometry

Related Career Pathways/Majors

Arts, A/V Technology & Communications Career Cluster

- A/V Technology & Film Career Pathway
- Telecommunications Career Pathway

Science, Technology, Engineering & Mathematics Career Cluster

- Engineering & Technology Pathway
- Science & Mathematics Pathway

Transferable Skills and Abilities

Concentration

- Designing and developing complex electrical systems and electronic components and products
- Keeping track of multiple design elements and technical characteristics when performing these tasks

Initiative

- Applying knowledge to new tasks in every project undertaken
- Engaging in continuing education to keep up with changes in technology

Interpersonal Skills

- Working with others during the manufacturing process to ensure that plans are implemented correctly
- Monitoring technicians and devising remedies to problems as they arise

Math Skills

- Using the principles of calculus and other advanced math to analyze, design, and troubleshoot equipment

Speaking Skills

- Working closely with other engineers and technicians
- Explaining designs and reasoning clearly and relaying instructions during product development and production
- Explaining complex issues to customers who have little or no technical expertise

Writing Skills

- Developing technical publications related to equipment they develop, including maintenance manuals, operation manuals, parts lists, product proposals, and design methods documents

Postsecondary

To enter the occupation, prospective electrical and electronics engineers need a bachelor's degree in electrical engineering, electronics engineering, electrical engineering technology, or a related engineering field. Programs include classroom, laboratory, and field studies. Courses include digital systems design, differential equations, and electrical circuit theory. Programs in electrical engineering, electronics engineering, or electrical engineering technology should be accredited by the Accreditation Board for Engineering and Technology, Inc. (ABET).

Some colleges and universities offer cooperative programs in which students gain practical experience while completing their education. Cooperative pro-

Fast Fact

Moore's Law, named after American engineer Gordon Moore, states that the number of transistors that can be packed onto a microchip doubles every two years, allowing for the exponential growth of processing power. For instance, microchips that could once host fewer than 100,000 transistors in the can today host more than 50 billion. Most speculators agree Moore's Law will end around 2025, although technologies such as machine learning and quantum computing will continue to evolve the field.

Source: newengineer.com

Photo via iStock/krystiannawrocki. [Used under license.]

grams combine classroom study with practical work. Internships provide similar experience and are growing in number.

At some universities, students can enroll in a 5-year program that leads to both a bachelor's degree and a master's degree. A graduate degree allows an engineer to work as an instructor at some universities, or in research and development.

Related College Majors

- Applied Physics
- Electrical, Electronics & Communications Engineering
- Engineering Mechanics
- Engineering Physics
- Engineering Science
- Environmental/Environmental Health Engineering
- Laser & Optical Engineering
- Materials Engineering
- Mechanical Engineering
- Telecommunications Engineering

Professional Certification and Licensure

Licensure is not required for entry-level positions as electrical and electronics engineers. A Professional Engineering (PE) license, which allows for higher levels of leadership and independence, can be acquired later in one's career. Licensed engineers are called professional engineers (PEs). A PE can oversee the work of other engineers, sign off on projects, and provide services directly to the public. State licensure generally requires:

- a degree from an ABET-accredited engineering program
- a passing score on the Fundamentals of Engineering (FE) exam
- relevant work experience, typically at least 4 years
- a passing score on the Professional Engineering (PE) exam

The initial Fundamentals of Engineering Examination (FE) can be taken after earning a bachelor's degree. Engineers who pass this exam commonly are called engineers in training (EITs) or engineer interns (EIs). After meeting work experience requirements, EITs and EIs can take the second exam, called the Principles and Practice of Engineering (PE).

Each state issues its own licenses. Most states recognize licensure from other states if the licensing state's requirements meet or exceed their own licensure requirements. Several states require continuing education for engineers to keep their licenses.

EARNINGS AND ADVANCEMENT

Earnings depend on industry, experience, and project scope and scale. Median annual earnings of electrical engineers were $100,830 in 2020. The lowest 10 percent earned less than $64,870, and the highest 10 percent earned more than $159,520. Median annual earnings of electronics engineers were $107,540 in 2020. The lowest 10 percent earned less than $69,210, and the highest 10 percent earned more than $167,410.

Electrical and electronics engineers may receive paid vacations, holidays, and sick days; life and health insurance; and retirement benefits. These are usually paid by the employer. Engineers who are required to travel may be able to recoup these expenses by their employer.

Electrical and electronic engineers may advance to supervisory positions in which they lead a team of engineers and technicians. Some may move to management positions, working as engineering or program managers. Preparation for managerial positions usually requires working under the guidance of a more experienced engineer.

For sales work, an engineering background enables engineers to discuss a product's technical aspects and assist in product planning and use.

EMPLOYMENT AND OUTLOOK

Electrical and electronic engineers held 313,200 jobs in 2020. Employment is expected to grow as fast as average for all occupations through the year 2030, at a rate of 7 percent. Employment growth is expected to be tempered by slow growth or decline in some industries, such as manufacturing and utilities.

Job growth for electrical and electronics engineers is projected to occur largely in professional, scientific, and technical services firms, as more companies are expected to tap the expertise of engineers for projects involving electronic devices and systems. These engineers also will be needed to develop sophisticated consumer electronics.

The rapid pace of technological innovation will create some demand for electrical and electronics engineers in research and development, an area in which engineering expertise will be needed to design distribution systems related to new technologies. These engineers will play key roles in new developments with solar arrays, semiconductors, and communications technologies.

Related Occupations

- Aerospace Engineer
- Architectural/Engineering Manager
- Bioengineer/Biomedical Engineer
- Computer Hardware Engineer
- Electrical/Electronics Engineering Technician
- Electrical/Electronics Installer/Repairer
- Electrician
- Electro-mechanical Technician
- Network/Computer Systems Administrator
- Sales Engineer

MORE INFORMATION

Accreditation Board for Engineering and Technology, Inc. (ABET)
415 North Charles Street
Baltimore, MD 21201
410.347.7700
www.abet.org

American Association of Engineering Societies (AAES)
1801 Alexander Bell Drive
Reston, VA 20191
202.296.2237
orders@aaes.org
www.aaes.org

American Society for Engineering Education (ASEE)
1818 N Street NW, Suite 600
Washington, DC 20036
202.331.3500
www.asee.org

Audio Engineering Society (AES)
132 East 43rd Street, Suite 405
New York, NY 10017
212.661.8528
www.aes.org

Engineering Education Service Center (EESC)
1411 Old Hardman Btms Road
Clarkesville, GA 30523
706.499.5011
www.engineeringedu.com

International Society of Automation (ISA)
67 T.W. Alexander Drive
Research Triangle Park, NC 27709
919.549.8411
www.isa.org

National Association of Power Engineers (NAPE)
1 Springfield Street, Suite 1
Chicopee, MA 01013
413.592.9273
www.powerengineers.com

National Council of Examiners for Engineering and Surveying (NCEES)
200 Verdae Boulevard
Greenville, SC 29607
800.250.3196
www.ncees.org

National Society of Professional Engineers (NSPE)
1420 King Street
Alexandria, VA 22314
888.285.6773
www.nspe.org

Technology Student Association (TSA)
1904 Association Drive
Reston, VA 20191-1540
703.860.9000
general@tsaweb.org
www.tsaweb.org

Stuart Paterson

Conversation With...
THOMAS SLOWE

Entrepreneur in Residence, Buffalo University
Buffalo, NY
Entrepreneur, 33 years

What was your individual career path in terms of education/training, entry-level job, or other significant opportunity?

Both of my parents were entrepreneurs, so I had a lot of guidance in terms of understanding the basic questions involved: what can you make, and how do you sell it? How do you identify what is going to be something people are interested in obtaining, what are they actually going to cough up money for, how much, and how much time, energy, and what materials do you need to produce that thing?

I started my first company when I was fourteen, and my family lived in Rochester, New York. At that time, neckties were conservative in style. I decided I wanted to make ties. My mom had been running a quilt-making business and had the sewing machines and fabric. My dad was involved in the arts community, and they had art shows. I used all sorts of flashy eighties and nineties colors and sold my completely wild ties at shows and made several thousand dollars.

I'm from a family of engineers and went on to earn my BS in electrical engineering and computer science from Rutgers University, then went straight to Massachusetts Institute of Technology and earned an MS in media arts and sciences from their Media Lab. This was during the dotcom boom of 1996-1998.

At Rutgers, I worked on making automated recognition systems to find things in video and images, and I graduated with published work. I also did a lot of artwork during school. I carried that research to MIT, where the Media Lab is all about the integration of art and science.

After graduating, I moved to Stockholm, Sweden, for 2 years to work for Ericsson Telecom, a company that had been a Media Lab sponsor. I enjoyed my time there and set up a lab for them but, overall, that experience validated that big companies were not for me.

I returned and got a job at a startup company called ObjectVideo in Reston, Virginia, outside Washington, DC. We worked with the intelligence community, and I was a senior scientist there for several years, although during this time I realized that I was not only interested in engineering. After a while I decided to start my own company and moved back to my hometown of Buffalo.

I've been involved with a number of startup companies, but Nervve was the biggest one that was cleanly founded by me. It centered on my experience working with the US intelligence community. I saw that they have and continue to acquire huge volumes of video and imagery. The published data point we had at the time showed that less than 2 percent of that video was ever seen by a human. The use of video for defense and intelligence was largely relegated to understanding what happened after the bomb already had gone off.

We essentially invented the core algorithm that made it possible to scan—very, very efficiently—a huge volume of video 1000 times faster than the competing technology at the time. We got investment from the intelligence community alongside a bunch of other people. We worked with foreign and domestic spy agencies, and then branched into sports media and worked with many major league teams. We were able to use the technology to measure the output of branding in an entire season's worth of games and combine that with viewership data to put a value on in-screen advertising. The company was sold in 2017.

I've always got five or six side hustles. I've built furniture. I've done designs for construction projects. I've got a few entrepreneurial ventures now with high hopes things will be solidified over the next 3 months. I'm also an Entrepreneur-in-Residence at Buffalo University and work with 10 to 12 companies—or embryonic ideas for companies—at any one time.

What are the most important skills and/or qualities for someone in your profession?
Risk tolerance, self-generated motivation, broad interests, and a willingness to constantly acquire new skills. You need an ability to withdraw from the situation from time to time to look at the big picture so you can validate whether what you're doing or who you are collaborating with makes sense. You need to be willing to try new things.

What do you wish you had known going into this profession?
I wish I had known to much more strictly judge individuals I do business with. As the stakes get higher, that tends to attract a higher percentage of bad players. My perceptions going in have been far more optimistic than they should have been because a lot of people are pretty Machiavellian.

Are there many job opportunities in your profession? In what specific areas?
There are an infinite number of job opportunities. It's all about you creating it and taking responsibility for it.

How do you see your profession changing in the next 5 years? How will technology impact that change, and what skills will be required?
I think that integration between traditionally siloed knowledge spaces is going to accelerate significantly. When I was a kid, lawyers did their thing, accountants did theirs. Now all that has changed. If you're a super geek who wants to go to engineering school, stretch yourself and do different things. Learn to cook. Take a class in fashion design.

What do you like most about your job? What do you like least about your job?

I love the feeling you get when you sell something you spent time working on. I love working with smart, motivated people. I really detest working with people who are not motivated and don't finish their work, or their work is of poor quality.

Can you suggest a "try this" for a student who may be considering a career in your field?

I would advise coming up with an idea and then figuring out what you can make or produce. That really is the crux of it. If you can get an internship or work with a small company, that's going to help you.

This interview was originally published in 2021.

Conversation With...

DANIEL W. BLISS

Associate Professor
Arizona State University, School of Electrical,
Computer & Energy Engineering
Systems Engineer, 30 years

What was your individual career path in terms of education/training, entry-level job, or other significant opportunity?

I enjoyed mathematics and physics and because my dad was a mechanical engineer, I was exposed to the idea of engineering. Because I was more math inclined, electrical engineering seemed the right path. I earned a B.S. in electrical engineering from Arizona State University.

As I looked around for what to do, rocket avionics seemed an interesting path and I took a job with General Dynamics in San Diego working on the Atlas-Centaur launch vehicle. Developing and designing rocket avionics is a lot of fun, but it is 90 percent testing. So, when my boss suggested a new opportunity to develop super conducting magnets for a high-energy particle accelerator, I took it. Basically, you take a particle—in my case, a proton—pass it in through an RF cavity that accelerates the particle and bend the path of the particle back to the cavity with strong magnets to accelerate it further. The particle accelerator we worked on was to understand the basic building blocks of the universe.

Concurrent with my job at General Dynamics, I began work toward an M.S. and a PhD in physics at the University of California, San Diego while doing much of my research at Cornell University. My research focused on high-energy particle physics.

As I finished my doctorate, unfortunately the U.S. particle accelerator project, the superconducting super collider (SSC)—similar to the one in Cern, Switzerland—shut down and the jobs dried up. I didn't want to go for a post-doctoral position because I saw really good people doing that for years and still not get a faculty position. My electromagnetics teaching assistant from grad school, who worked at the Massachusetts Institute of Technology, gave me a reference that helped me get a job at MIT's Lincoln Laboratory. It's a federally-funded national lab for technology development. I was there 15 years, first as technical staff and then as senior staff leading tens of millions of dollar programs and developing advanced communications and radar systems, and novel physiological signal processing.

I had always thought about being a professor and explored different possibilities. Just over five years ago, I found the position here at ASU. Ten PhD students and four research scientists work under me at the university's Bliss Laboratory for Information, Signals and Systems. We work on radar systems and communications systems for everything from your phone to the internet of things to control systems for various things that fly. Radar systems may be large-scale military systems or weather radars or extremely small-scale radars that measure cardiac signals. I also direct the Center for Wireless Information Systems and Computational Architectures. In addition to my research, I spend about a third of my time getting money to fund that research. I earned awards of more than $1.5 million in 2015-2016. I also teach a class every semester.

What are the most important skills and/or qualities for someone in your profession?
I view myself as a systems engineer who goes across the spectrum from fundamental theory to how does a piece of technology I built fit into the larger system? To do that, some skills are not surprising, such as a thorough understanding of mathematics, physics, statistics, and familiarity with the current state of technology.

However, you also need to be able to see the big picture. How does a technology fit into the user's needs and into the needs of society as a whole? You have to talk to users and sponsors to learn this. You need to be able to communicate your ideas, and, more importantly, listen and re-tune your ideas and goals based upon what they tell you. You need to work with a team, and manage the challenge of fitting projects to the skill sets of your team members.

Finally, there might be fallacy in the nature of the question. I think that the traditional idea of a profession will soon die. Because of technology, we are creating new and destroying old careers at a rate that we have never seen before. While the name of a profession may stay the same, the primary duties are becoming fluid. As a system engineer, I may be at the edge of this phenomenon, but to a greater or lesser extent, it will touch all professions soon. We all need to be looking for the next shift and to be constantly learning.

What do you wish you had known going into this profession?
Honestly, I don't think I have ever been really surprised by any aspects of my duties. However, I have been surprised when I observed occasional, although all too common, examples of sexism and racism. I have to admit that early in my career, I simply did not know how to respond. It is easy for me to tell people that it is unacceptable now, but I was not equipped for it then.

Are there many job opportunities in your profession? In what specific areas?
The opportunities for those who understand mathematics, physics, and computers are particularly rich. I believe many opportunities will arise from the implications of small-scale radars for a wide range of applications, such as individualized medical systems or human-machine gesture interfaces. Much like people thought that it would be ridiculous to put a camera on a phone, I think that many are underestimating the future opportunities of these tiny radars.

How do you see your profession changing in the next five years, how will technology impact that change, and what skills will be required?

Because we are the developers of technology, we need to spend more time thinking about user and societal implications. The fact is, all of our technologies will increasingly become integrated into every single aspect of life. As examples, we currently have a fundamental problem with our approach to the internet of things, or IoT, which is basically the practice of connecting devices to the internet and/or to each other. The way it is evolving, we are just putting more gadgets on the internet, but no one is really taking care of them. Maybe bad actors are taking over someone's Blu-ray player and using it as a bot. We are at the precipice of a wide range of privacy and security disasters associated with flawed system choices.

What do you enjoy most about your job? What do you enjoy least about your job?

I love coming up with system ideas that I get to explore mathematically and practically, and I enjoy writing about these ideas. On the less enjoyable side, I spend an inordinate amount of time working on contracting issues.

Can you suggest a valuable "try this" for students considering a career in your profession?

A friend told me that he likes to hire people who "live the lifestyle." He meant that he likes to hire people would do engineering projects at home for fun. For example, get GNU Radio running on your computer. Get a software-defined radio and see if you and a friend can communicate with a waveform of your own design. And, keep playing with math.

This interview was originally published in 2018.

Conversation With...
MIKE RUST

Electrical Engineer
Senior engineer/Co-owner, D2D Technologies, Jacksonville, FL
In the field, 35 years

What was your individual career path in terms of education/training, entry-level job, or other significant opportunity?
I knew I wanted a career in some type of science starting in high school. I like seeing things work, and I knew engineering had good-paying jobs, usually in high demand. I took a lot of math, physics, biology, and chemistry courses with this in mind. After high school, I started at the University of Maryland with a declared major of biology, but soon realized I really wanted to be an electrical engineer because it was a more financially lucrative field and there were more opportunities. I've always been practical like that. I earned a Bachelor of Science in electrical engineering and did all the coursework for my Master of Science in electrical engineering during the following two years. My first job was in avionics, designing circuit boards to allow systems on planes to communicate with each other. Over time, I moonlighted side jobs because I was always interested in having my own consulting business. When the company I worked for wanted to transfer me, I asked for a layoff because, at the time, they offered a good severance package. I took it and started my own business. I like consulting; I have more control over my time, am not tied to a commute, and can live where I want to. My specialty is PC (printed circuit) board design, which are the boards inside all of the electronics in your home and business. I also design field programmable gate arrays, which are used in broadcast facilities such as cable TV or satellite, as well as many other places.

What are the most important skills and/or qualities for someone in your profession?
When you're in school, math skills are needed and physics helps. Once you move into your professional life, the desire to make things work and work correctly all the time are key. You also need learn how to communicate technical subject matter to people who don't quite understand how something works.

What do you wish you had known going into this profession?
I wish I had known that it is difficult to remain a working engineer versus being pushed into management, but that is probably true with most professions. In engineering, about ten years into a career, people get moved into management. It's hard to find working engineers over 40 or 45; by then they've either washed out or moved into management. The technical stuff is

the fun stuff even though it's harder; in management, you're dealing with people and deadlines.

Are there many job opportunities in your profession? In what specific areas?
Yes, there are many opportunities in my field as it is a key part of so many things in today's world—smartphones, computers, robotics, the Internet, social media, space exploration, biomedical, telecommunications, solar energy, the automotive industry...

How do you see your profession changing in the next five years, what role will technology play in those changes, and what skills will be required?
Since electrical engineering is at the forefront of a lot of technological change, you must always stay on top of many of the latest computer technologies. Most of the tools we use to design products are computer-based software programs. Significant portions of the electrical engineering field are very close to computer science; sometimes we have to write the software to put our tools together.

Also, computer languages are always evolving and you need to stay abreast of that as well. I don't think the field will change drastically over the next few years, but it is constantly evolving and people in the field must be very adaptable.

What do you enjoy most about your job? What do you enjoy least about your job?
I really enjoy seeing something I created work as I envisioned it during the design process. I don't enjoy all the documentation—sometimes pages and pages—that must be done to ensure that someone else can reproduce the design, but it is necessary so other people can carry on my work since I'm a consultant.

Can you suggest a valuable "try this" for students considering a career in your profession?
Getting involved in programming languages and Arduino board projects and making something that works while you're still in high school (or before) are inexpensive ways to determine if you like doing that type of thing. For instance, you could start building a robot with an Arduino board and a couple of stepper motors. You could also write an app for a smartphone. Most people don't realize how close electrical engineering is to computer science these days, but an electrical engineer has many more career choices than a computer scientist.

This interview was originally published in 2015.

Electrical/Electronics Engineering Technologist/Technician

Snapshot

Career Cluster(s): Arts, A/V Technology & Communications; Science, Technology, Engineering & Mathematics

Interests: Electronics; Electrical Systems; Technology

Earnings (Yearly Average): $67,550

Employment & Outlook: Slower Than Average Growth Expected

OVERVIEW

Sphere of Work

Electrical and electronic engineering technologists and technicians help electrical and electronics engineers plan and develop communications equipment, computers, medical monitoring devices, or other equipment that is powered by other electricity or electric current. They often work in product evaluation and testing, using measuring and diagnostic devices to test, adjust, and repair equipment. They are also involved in assembling equipment for automation.

Photo via iStock/A stockphoto. [Used under license.]

Work Environment

Electrical and electronic engineering technologists and technicians may work in offices, laboratories, and factories, since their duties include both theoretical and assembly-line work. They most often work on teams.

Profile

Working Conditions: Inside and Outside
Physical Strength: Medium Work
Education Needs: Associate's Degree
Licensure/Certification: Optional
Opportunities for Experience: Internship; Co-op
Interest Score: RIC

Occupation Interest

Electrical and electronic engineering technologists and technicians are math- and science-oriented individuals who want to contribute to the creation of new electrical and electronic systems, components, and equipment. In many cases they aspire to become full electrical and electronic engineers who conceive and design these pieces of equipment.

Duties and Responsibilities

Electrical Engineering Technologist/ Technician

- Assembling electrical and electronic systems and prototypes
- Building, calibrating, and repairing electrical instruments or testing equipment
- Visiting sites where systems are made to observe conditions affecting design
- Identifying solutions to technical design problems that arise in making electrical systems
- Inspecting designs for quality control, report findings, and recommend changes, if necessary
- Drawing diagrams and write specifications about design details of experimental electronics units

Electronic Engineering Technologist/ Technician

- Creating basic circuitry and draft sketches to clarify details of design, under engineers' direction
- Building prototypes from plans or sketches
- Assembling, testing, and maintaining circuitry or electronic components according to engineering instructions, knowledge of electronics, and technical manuals
- Adjusting and replacing defective circuitry and electronic components
- Making parts, such as coils and terminal boards, using bench lathes, drills, or other machine tools

A Day in the Life—Duties and Responsibilities

Electrical engineering technologists and technicians install and maintain electrical control systems and equipment and adjust electrical prototypes, parts, and assemblies to correct problems. When testing systems, they set up equipment and evaluate how the parts, assemblies, or systems perform under simulated conditions. They also analyze test information to resolve design problems.

Electronic engineering technologists and technicians identify and repair equipment malfunctions. They also calibrate and perform preventive maintenance on equipment and systems.

These workers often need to read blueprints, diagrams, and engineering instructions for assembling electronic units. They also write reports and record data on testing techniques, laboratory equipment, and specifications.

WORK ENVIRONMENT

Immediate Physical Environment

Electrical and electronic engineering technologists and technicians work on teams with electrical and electronics engineers. They work in offices, laboratories, and factories because their job tasks involve both engineering theory and assembly-line production.

Electrical and electronic engineering technologists and technicians may be exposed to hazards from equipment or toxic materials, but incidents are rare if procedures are followed.

Most electrical and electronic engineering technologists and technicians work full time. Some work day or night shifts, depending on production schedules. In the federal government, their schedules usually follow a standard workweek.

Human Environment

Electrical and electronic engineering technologists and technicians frequently work on teams, so must be comfortable sharing ideas and collaborating in order to solve problems and accomplish tasks. They also report to senior engineers. In the field, they may be required to visit sites where their equipment will be utilized, so can expect to interact with employees at those locations, including troubleshooting problems.

Technological Environment

Computer-aided design (CAD) software is common, as is analytical or scientific software, development environment software, and object- or component-oriented development software. They may also be required to work with their hands and use analogue tools such as bench lathes, drills, and other machine tools.

EDUCATION, TRAINING, AND ADVANCEMENT

High School/Secondary

High school students interested in studying electrical or electronics engineering benefit from taking courses in physics and math, including algebra, trigonometry, and calculus. Courses in drafting are also helpful, because electrical and electronics engineers often are required to prepare technical drawings. During high school, students can attend engineering summer camps to see what these and other engineers do. Attending these camps can help students plan their coursework for the remainder of their time in high school. The Engineering Education Service Center (EESC) has a directory of engineering summer camps.

Suggested High School Subjects

- Algebra
- Biology
- Chemistry
- Civics
- Computer Science
- Drafting
- Earth or Environmental Science
- Economics
- English
- Entrepreneurship
- Geometry
- History
- Physics
- Pre-Calculus
- Psychology
- Statistics
- Trigonometry

Related Career Pathways/Majors

Arts, A/V Technology & Communications Career Cluster

- A/V Technology & Film Career Pathway
- Telecommunications Career Pathway

Science, Technology, Engineering & Mathematics Career Cluster

- Engineering & Technology Pathway
- Science & Mathematics Pathway

Transferable Skills and Abilities

Communication Skills

- Following instructions from engineers and others, as well as clearly conveying problems to engineers

Detail-oriented

- Paying attention to detail when assembling, troubleshooting, and repairing electronic and electrical mechanical systems

Math Skills

- Using mathematics for analysis, design, and troubleshooting tasks

Mechanical Skills

- Using hand tools and soldering irons on small circuitry and electronic parts to build components by hand

Problem-solving Skills

- Identifying and fixing problems that arise in assembling and inspecting electrical engineers' designs and prototypes

Writing Skills

- Writing reports about on-site construction, design problems, or testing results.
- Writing must be clear and well-organized to convey the information in the reports

Postsecondary

Associate's degree programs in electrical or electronic engineering technology are available at community colleges and vocational-technical schools. Programs accredited by the Accreditation Board for Engineering and Technology, Inc. (ABET) or other organizations typically include courses such as algebra, programming languages, physics, and circuitry.

Depending on-the-job tasks or the industry, employers may prefer to hire candidates who have a bachelor's degree. Candidates for other jobs may qualify with a high school diploma.

Related College Majors

- Applied Physics
- Electrical, Electronics & Communications Engineering
- Engineering Mechanics
- Engineering Physics
- Engineering Science
- Environmental/Environmental Health Engineering
- Laser & Optical Engineering
- Materials Engineering
- Mechanical Engineering
- Telecommunications Engineering

Professional Certification and Licensure

Certifications in a variety of fields are available for electrical and electronic engineering technologists and technicians. While optional, these credentials show that the designee has advanced knowledge. Among the organizations that offer certification are the National Institute for Certification in Engineering Technologies (NICET), Electronics Technicians Association (ETA) International, and the International Society of Automation (ISA).

EARNINGS AND ADVANCEMENT

Median annual earnings of electrical and electronic engineering technologists were $67,550 in 2020. The lowest 10 percent earned less than $40,170, and the highest 10 percent earned more than $99,810.

Electrical and electronic engineering technologists may receive paid vacations, holidays, and sick days; life and health insurance; and retirement benefits. These are usually paid by the employer. Engineers who are required to travel may be able to recoup these expenses by their employer.

EMPLOYMENT AND OUTLOOK

There were approximately 117,000 electrical and electronic engineering technologists and technicians employed nationally in 2020. Employment is projected to grow 2 percent from 2020 to 2030, slower than the average for all occupations.

Employment of these technologists and technicians in industries in which many are employed, such as manufacturing and federal government, is expected to decline. However, their employment is expected to grow in professional, scientific, and technical services firms as companies seek to contract out these services to lower costs.

Electrical and electronic engineering technologists and technicians also work closely with electrical and electronics engineers and computer hardware engineers in the computer systems design services industry. Demand is expected to be sustained by the continuing integration of computer and electronics systems, especially automation systems.

Related Occupations

- Aircraft/Avionics Equipment Mechanic/Technician
- Electrical/Electronics Engineer
- Electrical/Electronics Installer/Repairer
- Electromechanical/Mechatronics Technologist/Technician
- Mechanical Engineering Technologist/Technician

MORE INFORMATION

Accreditation Board for Engineering and Technology, Inc. (ABET)
415 N. Charles Street
Baltimore, MD 21201
410.347.7700
www.abet.org

American Association of Engineering Societies (AAES)
1801 Alexander Bell Drive
Reston, VA 20191
202.296.2237
orders@aaes.org
www.aaes.org

American Society for Engineering Education (ASEE)
1818 N Street, NW, Suite 600
Washington, DC 20036
202.331.3500
www.asee.org

Audio Engineering Society (AES)
132 E. 43rd Street, Suite 405
New York, NY 10017
212.661.8528
www.aes.org

Automation Federation
67 T.W. Alexander Drive, PO Box 12277
Research Triangle Park, NC 27709
919.549.8411
info@automationfederation.org
www.automationfederation.org

Electronics Technicians Association (ETA) International
5 Depot Street
Greencastle, IN 46135
800.288.3824
eta@etai.org
www.etai.org

Engineering Education Service Center (EESC)
1411 Old Hardman Btms Road
Clarkesville, GA 30523
706.499.5011
www.engineeringedu.com

International Society of Automation (ISA)
67 T.W. Alexander Drive
P.O. Box 12277
Research Triangle Park, NC 27709
919.549.8411
www.isa.org

National Association of Power Engineers (NAPE)
1 Springfield Street, Suite 1
Chicopee, MA 01013
413.592.9273
www.powerengineers.com

National Council of Examiners for Engineering and Surveying (NCEES)
200 Verdae Boulevard
Greenville, SC 29607
800.250.3196
www.ncees.org

National Institute for Certification in Engineering Technologies (NICET)
1420 King Street
Alexandria, VA 22314
888.476.4238
socoffice@soc.org
www.nicet.org

National Society of Professional Engineers (NSPE)
1420 King Street
Alexandria, VA 22314
888.285.6773
www.nspe.org

Technology Student Association (TSA)
1904 Association Drive
Reston, VA 20191-1540
703.860.9000
general@tsaweb.org
www.tsaweb.org

Stuart Paterson

Electromechanical/ Mechatronics Technologist/Technician

Snapshot

Career Cluster(s): Science, Technology, Engineering & Mathematics

Interests: Science; Technology; Mathematics; Engineering

Earnings (Yearly Average): $59,800

Employment & Outlook: Declining Growth Expected

OVERVIEW

Sphere of Work

Electromechnical and mechatronics technologists and technicians combine knowledge of mechanical technology with knowledge of electrical and electronic circuitry. They operate, test, and maintain unmanned, automated, robotic, or electromechanical equipment.

Work Environment

Electromechanical and mechatronics technologists and technicians work primarily indoors, both in offices and at production sites. Their work largely takes place in computer and electronic product manufacturing environments. They also deal in environments with the

Photo via iStock/hanohiki. [Used under license.]

production of larger machinery, and in professional, scientific, and technical services.

Profile

Working Conditions: Mainly Inside
Physical Strength: Light to Heavy Work (Varies)
Education Needs: Postsecondary Certificate; Associate's Degree
Licensure/Certification: Optional
Opportunities for Experience: Trainee; Entry-level Position
Interest Score: RIC

Occupation Interest

People interested in a career as an electromechanical and mechatronics technologist are technically minded people who work well under pressure. They enjoy working with electronics and have a fascination with how things work. Their interests will thus lead them to having knowledge of circuit boards, processors, chips, electronic equipment, and computer hardware and software, including applications and programming. They will also benefit from interest in and knowledge of machines and tools, including their designs, uses, repair, and maintenance.

Duties and Responsibilities

- Reading blueprints, schematics, and diagrams to determine the method and sequence of assembly of a machine or a piece of equipment
- Verifying dimensions of parts, using precision measuring instruments
- Operating metalworking machines to make housings, fittings, and fixtures
- Inspecting parts for surface defects
- Repairing and calibrating hydraulic and pneumatic assemblies
- Using instruments to test the performance of electromechanical assemblies
- Using soldering equipment and hand-tools to install electronic parts and hardware
- Operating, testing, or maintaining robotic equipment
- Analyzing and recording test results

A Day in the Life—Duties and Responsibilities

Electromechanical and mechatronics technologists and technicians test and operate machines in factories and at other worksites. They also document the tests they performed and analyze and record the results of those tests.

Electromechanical and mechatronics technologists and technicians install, maintain, and repair automated machinery and computer-controlled mechanical systems in industrial settings.

They also test, operate, or maintain robotic equipment at worksites. This equipment may include unmanned submarines, aircraft, or similar types of equipment for uses that include oil drilling, deep-ocean exploration, or hazardous-waste removal.

WORK ENVIRONMENT

Immediate Physical Environment

Electromechanical and mechatronics technologists and technicians work with electrical engineers and mechanical engineers. They work primarily in manufacturing industries, including those of computer and electronic products and of machinery, and in professional, scientific, and technical services. They often work at production sites and in offices.

Electromechanical and mechatronics technologists and technicians are sometimes exposed to hazards from equipment or toxic materials. However, incidents are rare as long as workers follow safety procedures.

Human Environment

Electromechanical and mechatronics technologists and technicians work with electrical engineers and mechanical engineers. They follow instructions laid out by these colleagues, and as such have many face-to-face interactions in the course of their work. The importance of being exact or accurate in their work is extremely high, so good listening skills and strong verbal skills are necessary and frequently employed.

Technological Environment

Electromechanical and mechatronics technologist and technicians possess a good deal of know-how in the realms of mechanics and electronics. These are the people who program robotic equipment and test the performance of equipment, as well as maintain that equipment. They must have knowledge of both a variety of software types (a variety of analytical, CAD, industrial control and enterprise resource planning software programs may be employed), and a range of industrial tools that they are able to use with precision.

EDUCATION, TRAINING, AND ADVANCEMENT

High School/Secondary

Students interested in electromechanical and mechatronics technology careers should focus on technical courses such as computer science. They might also consider joining their school's science club, if possible, for practical experience and relationship building.

Suggested High School Subjects
- Algebra
- Geometry
- Physics
- Pre-Calculus
- Trigonometry

Related Career Pathways/Majors
Science, Technology, Engineering & Mathematics Career Cluster
- Engineering & Technology Pathway

Postsecondary

Associate's degree programs and postsecondary certificates for electromechanical and mechatronics technologists and technicians are offered at vocational-technical schools and community colleges.

Employers may prefer to hire graduates of programs accredited by an organization such as the Accreditation Board for Engineering and Technology, Inc. (ABET). Associate's degree programs usually include courses in subjects such as algebra, trigonometry, and sciences. Depending on the program, students may have the option of concentrating in a field such as electromechanics, mechatronics, or industrial maintenance.

Related College Majors
- Electrical Engineering
- Mechanical Engineering

Transferable Skills and Abilities

Communication Skills
- Ability to follow instructions from engineers
- Conveying problems to engineers clearly

Detail-oriented
- Taking and recording the precise measurements that engineers need

Dexterity
- Using hand tools and soldering irons on small circuitry and electronic parts to create electronic components

Logical-thinking Skills
- Carrying out engineers' designs
- Inspecting designs for quality control
- Assembling prototypes
- Following a specific sequence or a set of rules

Math Skills
- Using mathematics for analysis, design, and troubleshooting in their tasks

Mechanical Skills
- Creating components for industrial machinery or equipment
- Operating equipment such as drill presses, grinders, and engine lathes

Problem-solving Skills
- Identifying and fixing problems that arise with engineering designs and prototypes

Writing Skills
- Writing clear, well-organized reports that describe on-site construction, testing results, and problems they found in carrying out designs

Adult Job Seekers

Adults interested in becoming an electromechanical and mechatronics technologist/technician, should enroll in either associate degree programs or postsecondary certificate programs available at community colleges and vocational schools. ABET-accredited programs are best, as many employers are looking for this on a potential employee's resume.

Professional Certification and Licensure

Electromechanical and mechatronics technologists and technicians may earn optional certification to demonstrate professional competence.

The International Society of Automation offers the Certified Control Systems Technician (CCST) and Certified Automation Professional (CAP) designations. Both require a written exam, and recertification is required after a specified number of years.

The National Institute for Certification in Engineering Technologies (NICET) offers certification in electrical power testing and other specialties. The technologist certification requires a four-year engineering technology degree.

EARNINGS AND ADVANCEMENT

The median annual wage for electromechanical and mechatronics technologists and technicians was $59,800 in 2020. The lowest 10 percent earned less than $37,350, and the highest 10 percent earned more than $95,340.

Electromechanical and mechatronics technologists and technicians may receive benefits such as paid vacations, holidays, and sick days; life and health insurance; and retirement benefits, usually paid by the employer.

Most electromechanical and mechatronics technologists and technicians work full time, and some work more than 40 hours per week.

EMPLOYMENT AND OUTLOOK

Employment of electromechanical and mechatronics technologists and technicians is projected to decline 2 percent from 2020 to 2030.

Despite declining employment, about 1,200 openings for electromechanical and mechatronics technologists and technicians are projected each year, on average, over the decade. All of those openings are expected to result from the need to replace workers who transfer to other occupations or exit the labor force, such as to retire.

Many of these technologists and technicians are employed in manufacturing industries, for which employment projections vary. Automation in manufacturing could affect this occupation in both positive and negative ways. While automation may replace certain responsibilities, electromechanical and mechatronics technologists and technicians will still be needed to operate and maintain the robotic equipment.

Related Occupations

- Drafter
- Electrical/Electronic Engineering Technologist/Technician
- Electrical/Electronics Engineer
- Electrical/Electronics Installer/Repairer
- Machinist/Tool & Die Maker
- Mechanical Engineer
- Mechanical Engineering Technologist/Technician

MORE INFORMATION

Accreditation Board for Engineering and Technology, Inc. (ABET)
415 N. Charles Street
Baltimore, MD 21201
410.347.7700
www.abet.org

American Society for Engineering Education (ASEE)
1818 N Street, NW, Suite 600
Washington, DC 20036
202.331.3500
www.asee.org

The American Society of Mechanical Engineers (ASME)
2 Park Avenue
New York, NY 10016-5990
800.843.2763
custcare@asme.org
www.asme.org

Automation Federation
67 T.W. Alexander Drive
PO Box 12277
Research Triangle Park, NC 27709
919.549.8411
info@automationfederation.org
www.automationfederation.org

Institute of Electrical and Electronics Engineers (IEEE)
3 Park Avenue, 17th Floor
New York, NY 10016-5997
800.678.4333
contactcenter@ieee.org
www.ieee.org

Technology Student Association (TSA)
1904 Association Drive
Reston, VA 20191-1540
703.860.9000
general@tsaweb.org
www.tsaweb.org

Daniel Vest

Environmental Engineer

Snapshot

Career Cluster(s): Science, Technology, Engineering & Mathematics

Interests: Science; Environmental Science; Engineering

Earnings (Yearly Average): $92,120

Employment & Outlook: Slower Than Average Growth Expected

OVERVIEW

Sphere of Work

Environmental engineers use the principles of engineering, soil science, biology, and chemistry to develop solutions to environmental problems. They work to improve recycling, waste disposal, public health, and water and air pollution control. They also address global issues, such as unsafe drinking water, climate change, and environmental sustainability.

Work Environment

Environmental engineers can work indoors or outdoors, depending on the job. They work in offices as a rule, but also can be required to work in the field, dealing with outdoor sites and helping assess environmental data. They also deliver their findings and communicate their know-how to people at general meetings and business conferences.

Photo via iStock/sturti. [Used under license.]

Occupation Interest

Environmental engineers are often investigative people, which means they are inquisitive and interested individuals who enjoy spending time alone with their thoughts. They are also traditional, which means they are detail-oriented and organized, and prefer to operate in a structured setting.

Profile

Working Conditions: Both Inside and Outside
Physical Strength: Light to Medium Work
Education Needs: Bachelor's Degree
Licensure/Certification: Optional
Opportunities for Experience: Co-op; Internship
Interest Score: IRC

A perhaps obvious requirement any environmental engineer should have: an interest in the environment, and the ways in which technology might improve overall conditions on the planet. Environmental engineers care deeply about the problems of reducing pollution, and creating sustainable living, and are constantly working towards these goals.

Thus, from as early an age as possible, they should have an aptitude in subjects such as chemistry, biology, physics, and math subjects like algebra, trigonometry, and calculus.

Duties and Responsibilities

- Preparing, reviewing, and updating environmental investigation reports
- Designing projects that lead to environmental protection, such as water reclamation facilities or air pollution control systems
- Obtaining, updating, and maintaining plans, permits, and standard operating procedures
- Providing technical support for environmental remediation projects and for legal actions
- Analyzing scientific data and doing quality-control checks
- Monitoring the progress of environmental improvement programs
- Inspecting industrial and municipal facilities and programs in order to ensure compliance with environmental regulations
- Advising corporations and government agencies about procedures for cleaning up contaminated sites

A Day in the Life—Duties and Responsibilities

Environmental engineers conduct hazardous-waste management studies in which they evaluate the significance of a hazard and advise on treating and containing it. They also design systems for municipal and industrial water supplies and industrial wastewater treatment, and research the environmental impact of proposed construction projects. Environmental engineers in government develop regulations to prevent mishaps.

Some environmental engineers study ways to minimize the effects of acid rain, climate change, automobile emissions, and ozone depletion. They also collaborate with environmental scientists, urban and regional planners, hazardous-waste technicians, and other engineers, as well as with specialists such as experts in law and business, to address environmental problems and environmental sustainability.

WORK ENVIRONMENT

Immediate Physical Environment

Environmental engineers work in a variety of settings because of the nature of the tasks they do:

- When they are working with other engineers and with urban and regional planners, environmental engineers are likely to be in offices.
- When they are working with businesspeople and lawyers, environmental engineers are likely to be at seminars, presenting information and answering questions.
- When they are working with hazardous materials removal workers and environmental scientists, environmental engineers work at specific sites outdoors.

Human Environment

Environmental engineers need to be good communicators, as their work can bring them into contact with other scientists, businesspeople, and lawyers, depending on the task at hand. They may be required to be on hand at specific outdoor sites, dealing with hazardous materials removal workers and environmental scientists. Environmental engineers often are asked to speak at seminars, presenting information to and answering questions for fellow scientists and laypeople alike.

Technological Environment

Environmental engineers will be familiar with and use a variety of technologies, including analytical/scientific software such as MATLAB and MODFLOW; they will be versed in different types of compliance software; computer-aided design (CAD) software, and more. In the course of their work, environmental engineers use tools such as: air sampling pumps; air velocity and temperature monitors; atomizers; bacteria transformation kits; conductivity meters, and filtering machinery.

EDUCATION, TRAINING, AND ADVANCEMENT

High School/Secondary

Students interested in becoming an environmental engineer should take high school courses in chemistry, biology, physics, and math, including algebra, trigonometry, and calculus.

Suggested High School Subjects

- Algebra
- Biology
- Calculus
- Earth or Life or Physical Science
- Geometry
- Physics
- Trigonometry

Related Career Pathways/Majors

Science, Technology, Engineering and Mathematics Career Cluster

- Engineering & Technology Pathway
- Science & Mathematics Pathway

Transferable Skills and Abilities

Imagination

- Designing systems that will be part of larger ones
- Foreseeing how the proposed designs will interact with components of the larger system, including the workers, machinery, and equipment, as well as with the environment

Interpersonal Skills

- Working with others toward a common goal
- Working with engineers and scientists who design other systems and with the technicians and mechanics who put the designs into practice

Problem-solving Skills

- Identifying and anticipating problems in order to prevent losses for their employers, safeguard workers' health, and mitigate environmental damage

Reading Skills

- Reading and understanding documents that deal with topics outside their scope of training, such as business and legal matters

Writing Skills

- Writing clearly so that others without their specific training can understand their documents, including plans, proposals, specifications, and findings, among others

Postsecondary

Entry-level environmental engineering jobs typically require a bachelor's degree in engineering. Programs include academic, laboratory, and fieldwork. Some college and university programs offer cooperative education in which students gain practical experience while completing their studies. Students also may choose to participate in internships.

At some colleges and universities, a student can enroll in a five-year program that leads to both a bachelor's and a master's degree. A graduate degree allows an engineer to work as an instructor at some colleges and universities or to do research and development (R&D), and some employers prefer to hire candidates who have a master's degree.

Related College Majors

- Agricultural Engineering
- Biomedical Engineering
- Environmental Science
- Geological Engineering
- Geoscience Engineering
- Ocean Engineering

Adult Job Seekers

Adults interested in going into environmental engineering will find that a bachelor's degree in environmental engineering or a related field, such as civil, chemical, or general engineering, is necessary. Schools that offer cooperative programs or internships are preferred by many employers.

Professional Certification and Licensure

Licensure is not required for entry-level positions as an environmental engineer. A Professional Engineering (PE) license, which allows for higher levels of leadership and independence, can be acquired later in one's career. After licensing, environmental engineers can earn board certification from the American Academy of Environmental Engineers & Scientists (AAEES). This certification shows that an environmental engineer has expertise in one or more areas of specialization.

EARNINGS AND ADVANCEMENT

The median annual wage for environmental engineers in 2020 was $92,120. The lowest 10 percent earned less than $55,450, and the highest 10 percent earned more than $144,670.

Environmental engineers may receive benefits such as paid vacations, holidays, and sick days; life and health insurance; and retirement benefits, usually paid by the employer.

EMPLOYMENT AND OUTLOOK

Employment of environmental engineers is projected to grow 4 percent from 2020 to 2030, slower than the average for all occupations.

State and local governments' concerns about water are leading to efforts to increase the efficiency of water use. Such a focus differs from that of wastewater treatment, for which this occupation is traditionally known. Most of the projected employment growth for environmental engineers is in professional, scientific, and technical services, as governments at the state and local levels draw on the industry to help address water efficiency concerns.

The federal government's requirements to clean up contaminated sites are expected to help sustain demand for these engineers' services. In addition, wastewater treatment is becoming a larger concern in areas of the country where drilling for shale gas requires the use and disposal of massive volumes of water.

Environmental engineers should continue to be needed to help utility companies and water treatment plants comply with federal or state environmental regulations, such as regulations regarding emissions from coal-fired power plants.

Related Occupations

- Chemical Engineer
- Civil Engineer
- Environmental Engineering Technologist/Technician
- Environmental Scientist/Specialist
- Hydrologist
- Natural Sciences Manager

MORE INFORMATION

American Academy of Environmental Engineers & Scientists (AAEES)
147 Old Solomons Island Road, Suite 303
Annapolis, MD 21401
410.266.3311
info@aaees.org
www.aaees.org

American Society for Engineering Education (ASEE)
1818 N Street, NW, Suite 600
Washington, DC 20036
202.331.3500
www.asee.org

Association of Environmental Engineering & Science Professors (AEESP)
1211 Connecticut Avenue, NW, Suite 650
Washington, DC 20036
202.640.6591
www.aeesp.org

Engineers for a Sustainable World (ESW)
424 Sapphire Way
Pittsburgh, PA 15224
412.475.9531
info@eswusa.org
www.eswglobal.org

Environmental and Engineering Geophysical Society (EEGS)
1391 Speer Boulevard, Suite 450
Denver, CO 80204
303.531.7517
http://www.eegs.org

Daniel Vest

Conversation With...
MARGUERITE DUFFY

Senior positions, Environmental Protection Agency
Washington, DC
Retired, 25 years in field

What was your individual career path in terms of education/training, entry-level job, or other significant opportunity?

I grew up in Washington, DC, and on my side of town there were bulldozers everywhere. I was watching the trees and forest in the city disappear to development, and it was something I looked at sadly. I always wondered how I could stop it, preserve it and protect it.

I had to work when I got out of high school, so I worked for the old federal Department of Health, Education and Welfare in Washington. Then I worked at the EPA's Chesapeake Bay Program as I earned my AB in English and American Studies at Goucher College. I minored in environmental studies. Many years later, I earned my Master's in Public Administration from American University.

During those college years, I took class credits by writing about the work scientists were doing on research boats. That experience cemented that I love writing and I could be useful in this field by communicating the work the scientists do.

Over my career, I held many senior positions at EPA. I took on the environmental correspondence for President George H.W. Bush. I managed the advance, briefing and scheduling office for the EPA Administrator. I worked as a senior advisor in the Chesapeake Bay Program. I was the Deputy Director for Communication and Outreach at Coastal America, a former multi-agency office aimed at protecting the nation's coastline.

We need everybody to understand science, and we need to put it in terms the public can understand. At the bay program, I developed fact sheets based on science. At one point, I was the senior editor for the enforcement and compliance office at EPA, and every document came through my desk so I could approve it or edit it for publication. The public needs to understand the law.

After many years working at the federal level, I retired to commit more time to local issues.

What are the most important skills and/or qualities for someone in your profession?
Know the science, the laws, the players and how to effectively communicate. Ask questions. Be resilient and adapt as funding and priorities change, as the work you do may shift. Right now I believe we're moving beyond protection and into an era of preparedness, readying for the big climate change disasters: droughts, hurricanes, floods, fires, viruses. The science predicted is not if these things will happen but when.

What do you wish you had known going into this profession?
My experience in Washington DC was at the national level. I wish I had gotten experience at the local level of government and more grassroots organizations for a broader understanding and appreciation of what others do and need. Though I loved my work in the policy area, I believe broader exposure would have accelerated my understanding.

Are there many job opportunities in your profession? In what specific areas?
Professionals are needed at all levels: science, technology and communication. Though government funding has lessened today, the needs have not gone away. It will not stay this way forever. In addition to government jobs, consider non-government organizations, fund raising, and companies that must comply with environmental standards. Students at all grade levels need ecology teachers. Even financial markets need environmental expertise to prevent environmental risk.

How do you see your profession changing in the next five years, how will technology shift, and what skills will be required?
The viruses that are out there, the way we grow food, the way we raise animals and the diseases that are caused—how do we shift that for the future? How do we shift away from burning forests to make way for farms? Or the extreme weather that's coming upon us? How do you prepare for these? It's not the if, it's the when. Until things become personal, like it has with the pandemic, people don't seem to react. I feel like environmental protection specialists are going to become environmental preparedness specialists.

What do you enjoy most about your job? What do you enjoy least about your job?
I love the constant learning. I was constantly learning new research, new facts, new ideas. I'm an avid reader and my environmental work suited my liberal arts tendencies. What I didn't like was too much bureaucracy and bean counting, but it's also a necessity.

Can you suggest a valuable "try this" for students considering a career in your profession?
Certainly, intern in your field of study. But if you are a scientist, you should try your hand at interning at a TV or radio outlet so you learn the importance of communicating the writing you are studying. If you're a journalist, intern in a laboratory to understand the science. It's also important for a scientist to understand the politics and policy; take a job or internship in the mayor's office. Environmental work is multi-faceted.

This interview was originally published in 2020.

Conversation With...
JAN VERTEFEUILLE

Senior Director
World Wildlife Fund
Advocacy, 15 years

What was your individual career path in terms of education/training, entry-level job, or other significant opportunity?
I transitioned into international advocacy for conservation after working at newspapers and loving it. But after a decade, I started thinking about what else I could do with my strongest skills: researching, writing, delivering work on deadline and drawing stories and information out of people.

I went to work for a nonprofit in D.C. called Environmental Media Services, which was run by a bunch of former reporters who educated journalists about environmental issues and worked to generate media coverage of issues like environmental health and genetically engineered food. I worked there for four years before learning of a communications job at World Wildlife Fund. Since my passion was wildlife, it seemed a great fit.

My job focuses on figuring out advocacy strategies to achieve a specific wildlife conservation goal, like trying to reduce illegal ivory sales in a particular country to reduce elephant poaching. We might launch a campaign that includes policy advocacy (i.e. lobbying), media outreach to raise public awareness, and direct public engagement, like asking citizens to pledge not to buy ivory or sign a petition to their government. Then we look at who might be influential with the decision-makers in that country. Would diplomatic pressure help? Would a celebrity or influential business leader take up the cause?

What are the most important skills and/or qualities for someone in your profession, particularly someone who decides to work overseas?
People think of advocacy as lobbying legislators and government officials to get a law passed, and that's part of it. But to be successful, you need to really understand the players you're trying to influence and what tactics will be successful and what ones might actually hurt your efforts.

Professional training and a degree is less important than innate skills, which can be cultivated and improved through practice: creativity, curiosity, strategic thinking, flexibility, persuasion and the ability to work with a cross-cultural team. And by that last one I mean the ability to really listen to members of your team from other cultures and draw out their ideas. During a brainstorming discussion, for instance, Americans tend to be forthright with their ideas and have no problem sharing ideas. Colleagues from other cultures may be more reluctant to speak up. I've learned the hard way that some people I work with find it very rude to tell me my idea is a bad one even when they know it won't work, so I now ask a lot of questions and listen carefully to the answers. I then offer ideas in ways that can be rejected by someone without the fear of being rude.

What do you wish you had known before deciding to work abroad?

I wish I'd known how fulfilling international work is—I would have started sooner! I love working with people all over the world even though it means holding conference calls at crazy hours, so that team members in Asia and North America can talk to each other. I also wish Americans were better informed about issues outside the U.S. There is a whole big fascinating world full of people who are just as smart, well-educated, compassionate and creative, and they're working just as hard to protect the environment where they live and improve their quality of life. It seems obvious, but it's worth stating explicitly.

Are there many job opportunities overseas in your profession? In what specific geographic areas?

As the population hurtles close to 9 billion people on the planet, there will be a surge in need for people interested in advocating for change internationally, whether it's figuring out how to feed all those people while protecting the environment, mitigating climate change before our coasts are swamped, ensuring elephants and tigers continue to roam in the wild, or improving the way we build cities. Nonprofit organizations and governments in wealthy countries like the U.S. that have a long history of international charitable giving will be doing a lot of the hiring.

Will the need for professionals in your career to travel and live overseas change in the next five years? What role will technology play in those changes, and what skills will be required?

Advances in video conferencing and communication apps like Skype, Webex, Whatsapp and Line have provided huge benefits in being able to keep in touch with colleagues all over the world at all hours of the day. That has cut down on the need for some international travel. But there is really no substitute for face-to- face interaction with colleagues, so I think some international travel will always be a part of jobs like mine.

Social media and platforms like change.org, Avaaz and other petition sites have increased opportunities to bring global attention to a problem. A great example is when WWF launched a campaign to press the president of Mexico to take more action to save the world's most endangered marine mammal, the tiny vaquita porpoise that's found only in Mexico's Gulf of California. We generated hundreds of thousands of emails to his office and launched a social media effort that Leo DiCaprio joined on Twitter. The president of Mexico, WWF and DiCaprio started a public dialogue on Twitter about the vaquita and the president and DiCaprio ended up signing a formal agreement to work together to save this critically endangered species.

What do you enjoy most about your job? What do you enjoy least about your job?
I enjoy being part of a global organization with such a diverse number of issues to work on (endangered species, climate change, fresh water, forests and oceans) and with team members all over the world to learn from.

My least favorite part of the job is dealing with budgets. Funding and tracking spending is important, but not something I enjoy.

Can you suggest a valuable "try this" for students considering a career overseas in your profession?
Start by writing a letter to the editor of your local paper or emailing the appropriate decision-makers about an issue you care about (school funding, lack of bike lanes in town, not enough recycling in the parks). Be constructive (criticizing people immediately makes them less likely to listen to you) and consider the issue from their perspective so you can be persuasive, and propose a solution. Will it require a new rule, for instance, to stop selling plastic water bottles at park concessions? What can you suggest to make that feasible?

This interview was originally published in 2017.

Conversation With...

CHRISTOPHER L. OVERCASH, P.E., BCEE, LEED AP

Senior Associate
KCI Technologies, Inc., MD
Environmental Engineer, 17 years

What was your individual career path in terms of education/training, entry-level job, or other significant opportunity?

When I went to college, I started in mechanical engineering. Like all engineering majors, I took core courses for the first two years. I changed to environmental engineering after I took a cross-country trip with a friend who later became a Navy pilot. We flew all the way across the country and back in a small Cessna. I always had a love of the outdoors, and we got to see all of the landscapes across the entire country. That flipped the switch; I decided to become an environmental engineer so I could participate in affecting and preserving those landscapes. After I graduated from the University of Maryland, I started working in a branch of environmental engineering called industrial hygiene. I didn't really like the focus of that work so I got a job that dealt initially with hazardous waste assessment. As I progressed in my technical knowledge this broadened into water and wastewater work as well as marine environmental projects.

When I graduated, I had hoped to get a job that paid for me to get my master's degree and fortunately, my employer did that. I got my Master's in Environmental Engineering at Johns Hopkins University, and, in addition to my day-job, I teach environmental engineering part-time through Johns Hopkins' Engineering for Professionals program.

What are the most important skills and/or qualities for someone in your profession?

The key is to have that base education you get in school—mathematics and science—so you are able to think like an engineer. You have to be able to take those tools and put them to work to solve all sorts of different problems. Even though you are working in a certain track of engineering, you may need to bring in knowledge of another track. That's a plus of being an environmental engineer: it's a broad-based field, so you are more effective leading a design team. Also, you need to be able to think creatively to solve problems.

What do you wish you had known going into this profession?

One thing you don't learn as an undergraduate is how to write to convey information. About ten years into your career, you're likely to be managing a team and you need to be able to convey information well in a written format.

Are there many job opportunities in your profession? In what specific areas?

For environmental engineers, there are many opportunities right now in water-wastewater because the infrastructure of our original water and sewer systems in the U.S. are coming to the end of their life expectancy. In many large cities it's failing at a significant rate, and it's absolutely necessary for the continued strength of the U.S. economy that this be replaced and fixed. You've got to be able to flush toilets and have drinking water. Other branches of environmental engineering that continue to be strong—although maybe not as strong—are remediation of hazardous waste sites and air pollution control.

How do you see your profession changing in the next five years, what role will technology play in those changes, and what skills will be required?

We've gone from pencils for drafting to CAD (computer assisted design), and CAD has continued to evolve such that design is becoming more and more automated. It's hard to say what leaps and bounds may happen in the next five years. CAD really drives efficiency for a project team; many junior engineers are filling the drafting role with CAD.

What do you like most about your job? What do you like least about your job?

The thing I enjoy most is the variety of projects I get to work on. I do water-wastewater projects, marine-environmental projects such as dredging, and led the design of a visitor's dock at Thomas Point Lighthouse in the Chesapeake Bay. I designed a dredged material containment system to re-create a wetland-system on Grand Lake in Ohio. I also have completed projects to assess hazardous materials such as going to NASA in Bermuda to look at what materials were left there that could be contaminating the soil.

Today I spend more time in the office than I would like now that I'm in a management role. When you're just coming up in the ranks, you spend a good portion of your time out in the field assessing project sites.

Can you suggest a valuable "try this" for students considering a career in your profession?

Internships and shadowing a professional are key to figuring out what an engineer does. They are very different from building base knowledge as you do in college.

This interview was originally published in 2014.

Conversation With...
JANE MADDEN

Senior Vice President
CDM Smith

What was your career path?

When I was young I was captivated by The Undersea World of Jacques Cousteau and convinced that I wanted to be a marine biologist so I too could spend my life appreciating the majestic beauty of this hidden world. I entered the University of Vermont as a biology major, but realized after my first year that this was not a perfect fit. Having always excelled in math, I was encouraged to consider engineering. I transferred to civil engineering, and through that curriculum discovered environmental engineering. After graduation, I immediately got a great job with CDM Smith, a consulting, engineering, construction, and operations firm, and have been with the same company for nearly thirty years, protecting the rivers and oceans I love through my work in the design of wastewater treatment facilities.

What are three pieces of advice?

1. Find an occupation that you are passionate about—look for a career, not a job.

2. Work with people you like, with similar core values.

3. Challenge those you work for—question authority; it is these questions that make people think differently, and advance society. Lastly, don't pass up an opportunity when it is offered to you, even though you think you may not be capable—if someone has faith in you, have faith in yourself.

What paths for career advancement are available to you?

The sky is the limit. I have grown up in my current organization starting out as a project engineer, moving to a project manager, then a client manager responsible for sales, and now am responsible for all the public sector work we do in Massachusetts. Since we are a global firm, opportunities exist for me around the world to assist in bringing clean water and sanitation to the global population. With environmental concerns related to greenhouse gas emissions and climate change, the focus of our industry has moved to the water-energy nexus and the vulnerability of communities given rising sea levels and more extreme droughts and rain events. The next step for me is being more involved in policymaking at a state and federal level to ensure that the government is making the right decisions with respect to the environment.

This interview was originally published in 2013.

Environmental Engineering Technologist/Technician

Snapshot

Career Cluster(s): Science, Technology, Engineering & Mathematics

Interests: Science; Earth Science; Environmental Health; Engineering

Earnings (Yearly Average): $51,630

Employment & Outlook: Average Growth Expected

OVERVIEW

Sphere of Work

Environmental engineering technologists and technicians carry out the plans that environmental engineers develop. This can involve work in laboratories, facilities, and in the field.

Work Environment

Photo via iStock/DSCimage. [Used under license.]

Environmental engineering technologists and technicians work in a variety of locations, depending on the assignment: they work in laboratories, as well as in fieldwork, collecting samples such as soil samples or groundwater samples. Sometimes their work requires them on a jobsite, inspecting the facilities and dealing with issues such as equipment set-up.

Profile

Working Conditions: Both Inside and Outside
Physical Strength: Light to Medium Work
Education Needs: Associate's Degree
Licensure/Certification: Varies by State
Opportunities for Experience: Trainee; Entry-level positions
Interest Score: RIC

Occupation Interest

People who pursue careers as environmental engineering technologists and technicians tend to be realistic types who enjoy work that deals with practical, hands-on problems and solutions. Investigative and conventional in nature, they may be attracted to this career as it involves searching for facts and figuring out problems mentally. They may prefer working with data and details more than with ideas and they enjoy carrying out plans developed by environmental engineers. The best people suited for this role display great initiative and are responsible types who rise to the challenges presented by the work.

A Day in the Life—Duties and Responsibilities

Environmental engineering technologists and technicians work both indoors and outdoors help to ensure environmental quality. Their tasks aid environmental engineers in developing solutions to control, prevent, and mitigate damage caused by pollution and other environmental problems.

In laboratories, environmental engineering technologists and technicians record observations and test results and document photographs. To keep laboratories supplied, they also may gather product information, identify vendors and suppliers, and order materials and equipment.

Duties and Responsibilities

- Setting up, testing, operating, and modifying equipment used to prevent or clean up environmental pollution
- Maintaining project records and computer program files
- Collecting and analyzing samples, such as of groundwater, for monitoring pollution or treatment
- Reviewing documents to ensure that they are complete and conform to reporting requirements
- Reviewing work plans to schedule activities
- Arranging for the disposal of asbestos, lead, and other hazardous materials

In the field, environmental engineering technologists and technicians may collect air, soil, or ground and surface water samples of an area. They also prepare and clean equipment, operate field sampling pumps, and monitor instruments.

They also may inspect facilities for compliance with regulations governing substances such as asbestos, lead, and wastewater.

WORK ENVIRONMENT

Immediate Physical Environment

Environmental engineering technologists and technicians work under the direction of environmental engineers and as part of a team with other technicians. Environmental engineering technologists and technicians may work indoors, often in laboratories. They also work outdoors, sometimes in remote locations.

Because environmental engineering technologists and technicians help with environmental cleanup, they may be exposed to hazards from equipment, chemicals, or toxic materials. For this reason, they must follow proper safety procedures, such as wearing hazmat suits and respirators as needed. When they work in wet areas, environmental engineering technologists and technicians wear rubber boots to keep their legs and feet dry.

Human Environment

Environmental engineering technologists and technicians follow the plans laid out by and thus work closely with environmental engineers; therefore reading, understanding, and executing instructions is essential. The technologists and technicians in this job will sometimes work as a team; communication and cooperation are key to do good work. They will also have to work with management at the sites they are working, where again communication is very important as they may need to explain in professional and friendly terms what they're doing as they install, test, and modify equipment at these sites.

Technological Environment

The technological requirements of environmental engineering technologists and technicians are varied, depending on the job. Their work is frequently hands-on in nature and their tools range from air quality monitoring and water quality testing instruments; to computers; groundwater monitoring equipment; oil-in-water monitors; radiation detectors; volatile organic compound analyzers; and industrial equipment.

EDUCATION, TRAINING, AND ADVANCEMENT

High School/Secondary

Prospective environmental engineering technologists and technicians should take science and math courses in high school to prepare for postsecondary programs in engineering technology.

Suggested High School Subjects

- Algebra
- Biology
- Calculus
- Earth or Life or Physical Science
- Geometry
- Physics
- Trigonometry

Related Career Pathways/Majors

Science, Technology, Engineering & Mathematics Career Cluster

- Engineering & Technology Pathway

Transferable Skills and Abilities

Communication Skills

- Listening attentively and conveying information to others

Critical-thinking Skills

- Helping identify problems and solutions and implementing the engineers' plans

Observational Skills

- Evaluating situations, recognizing problems, and informing environmental engineers as quickly as possible

Problem-solving Skills

- Implementing plans designed by environmental engineers
- Resolving issues that arise, such as unexpected findings during fieldwork

Reading Skills

- Understanding legal and technical documents in order to ensure that regulations are being met

Postsecondary

Employers usually prefer to hire candidates who have completed the Accreditation Board for Engineering and Technology, Inc. (ABET)-accredited postsecondary programs. Although some candidates may be hired with a high school diploma and postsecondary coursework, environmental engineering technologists and technicians typically need an associate's degree in environmental engineering technology or a related field to enter the occupation.

Associate degree programs in environmental engineering technology are available in community colleges and vocational-technical schools. These programs generally include courses in chemistry, environmental assessment, hazardous-waste management, and mathematics. Some environmental engineering technologists and technicians enter the occupation with a bachelor's degree.

Related College Majors

- Chemical Engineering

- Environmental Health & Safety
- Health Sciences
- Industrial Engineering
- Mechanical Engineering
- Systems Engineering

Adult Job Seekers

Adults interested in going into environmental engineering technologist/technician work would do best to acquire an associate's degree in environmental engineering technology or a related field. These programs are available through community colleges and vocational-technical schools. Without this type of degree, adults with bachelor's degrees in related science courses may have the best chance of breaking in.

Professional Certification and Licensure

Some states require environmental technologists and technicians to have permits or licenses to remove hazardous waste. Workers also may be required to have Occupational Safety & Health Administration (OSHA) Hazardous Waste Operations and Emergency Response Standard (HAZWOPER) certification. HAZWOPER certification includes training in health hazards, personal protective equipment, site safety, recognizing and identifying hazards, and decontamination. Refresher training may be required to maintain certification.

Additional Requirements

Most environmental engineering technologists and technicians work full time and typically have regular hours. However, they may work irregular hours in order to monitor operations or contain a major environmental threat.

EARNINGS AND ADVANCEMENT

The median annual wage for environmental engineering technologists and technicians was $51,630 in 2020. The lowest 10 percent earned less than $32,850, and the highest 10 percent earned more than $87,730.

Agricultural engineers may receive benefits such as paid vacations, holidays, and sick days; life and health insurance; and retirement benefits, usually paid by the employer.

Environmental engineering technologists and technicians usually begin work as trainees in entry-level positions and are supervised by an environmental engineer or experienced technician. As they gain experience, technologists and technicians take on more responsibility. Some advance to become senior environmental technologists and technicians or lead environmental technologists and technicians, functioning as supervisors on-site.

Technicians with a bachelor's degree may advance to become environmental engineers.

EMPLOYMENT AND OUTLOOK

Employment of environmental engineering technologists and technicians is projected to grow 8 percent from 2020 to 2030, about as fast as the average for all occupations.

Employment in this occupation is typically tied to projects created by environmental engineers. State and local governments are expected to focus their efforts and resources on efficient water use, stormwater management, and wastewater treatment over the next decade. The demand for more environmental technologists and technicians by consulting firms will arise as governments and larger firms look to reduce costs.

Related Occupations

- Environmental Engineer
- Environmental Science/Protection Technician
- Environmental Scientist/Specialist
- Hazardous Materials Removal Worker

MORE INFORMATION

Accreditation Board for Engineering and Technology, Inc. (ABET)
415 N. Charles Street
Baltimore, MD 21201
410.347.7700
comms@abet.org
www.abet.org

Occupational Safety and Health Administration (OSHA)
200 Constitution Avenue, NW
Washington, DC 20210
800.321.6742
www.osha.gov

Technology Student Association (TSA)
1904 Association Drive
Reston, VA 20191-1540
703.860.9000
general@tsaweb.org
tsaweb.org

Daniel Vest

Health & Safety Engineer

Snapshot

Career Cluster(s): Science, Technology, Engineering & Mathematics

Interests: Science; Earth Science; Environmental Health; Engineering

Earnings (Yearly Average): $94,240

Employment & Outlook: Average Growth Expected

OVERVIEW

Sphere of Work

Health and safety engineers develop procedures and design systems to protect people from illness and injury and property from damage. They combine knowledge of engineering and of health and safety to make sure that chemicals, machinery, software, furniture, and other products will not cause harm to people or damage to property.

Work Environment

Health and safety engineers typically work in offices. However, they also must spend time at worksites when necessary, which sometimes requires travel.

Photo via iStock/af_istocker. [Used under license.]

Occupation Interest

Health and safety engineers are both technically minded and compassionate in nature. While their work is scientific, it is also highly people-oriented. They are frequently thinking about and designing systems intended to better the lives of their fellow man. Keeping everyone safe from harm is the primary goal of any health and safety engineer, and they must accordingly display forethought as they anticipate the worst scenarios and come up with ways of avoiding bad outcomes. An attention to fine technical detail is a necessity; education-wise, an aptitude for math and science courses is a must.

Profile

Working Conditions: Both Inside and Outside
Physical Strength: Light to Medium Work
Education Needs: Bachelor's Degree
Licensure/Certification: Optional
Opportunities for Experience: Co-op; Internship
Interest Score: IRC

Duties and Responsibilities

- Maintaining and applying knowledge of current health and safety policies, regulations, and industrial processes
- Reviewing plans and specifications for new machinery and equipment to make sure that they meet safety requirements
- Identifying and correcting potential hazards by inspecting facilities, machinery, and safety equipment
- Evaluating the effectiveness of various industrial control mechanisms
- Ensuring that buildings or products comply with health and safety regulations, especially after an inspection that required changes
- Installing safety devices on machinery or directing the installation of these devices
- Reviewing employee safety programs and recommending improvements

A Day in the Life—Duties and Responsibilities

In addition to their work developing procedures and systems designed to keep people safe, health and safety engineers also investigate industrial accidents and injuries to determine their causes and to determine whether the incidents were avoidable or can be prevented in the future. They interview employers and employees to learn about work environments and incidents that lead to accidents or injuries. They also evaluate the corrections that were made to remedy violations found during health inspections.

OCCUPATION SPECIALTIES

Fire Prevention and Protection Engineer

Fire prevention and protection engineers conduct analyses and make recommendations regarding the potential fire hazards of buildings, materials, and transportation systems. They also design, install, and maintain fire prevention and suppression systems and inspect systems to ensure that they meet government safety regulations. Fire prevention and protection engineers must be licensed and must keep up with changes in fire codes and regulations.

Product Safety Engineer
Product safety engineers, sometimes called product compliance engineers, develop and conduct tests to make sure that various products are safe and comply with industry or government safety regulations. These engineers work on a wide range of products, from nuclear submarine reactors and robotics to cell phones and computer systems.

System Safety Engineer
Systems safety engineers identify and analyze risks and hazards associated with system designs in order to make them safe while ensuring that the systems remain operational and effective. They work in many fields, including aerospace, and are moving into new fields, such as software safety, medical safety, and environmental safety.

WORK ENVIRONMENT

Immediate Physical Environment

Health and safety engineers work in a variety of locations. They can spend a good deal of time in office environments, developing plans. They are also required at times to leave the office, for instance when they need to investigate industrial accidents and injuries to determine their causes and come up with the answers as to why it happened, if it will happen again in the future, and how to prevent it.

Human Environment

Health and safety engineers can work solo when it comes to designing plans and systems, and a good deal of their time working is indeed spent in contemplative office work; but they must also frequently visit different worksites to observe and evaluate the health and safety conditions there. Talking with and gaining information from owners, supervisors, and workers alike, it is important for them to have strong powers of communication.

Technological Environment

The technological requirements of health and safety engineers are varied, depending on the job. They must frequently visit worksites, inspecting the facilities and coming away from that experience able to identify and correct any potential hazards. This can involve reviewing plans and specifications for new machinery and equipment to make sure that they meet safety requirements; they may also need to install or direct the installation of safety devices on preexisting machinery.

EDUCATION, TRAINING, AND ADVANCEMENT

High School/Secondary

Students interested in health and safety engineering careers should focus on science and math courses, as well as technical courses such as computer science. They should also consider joining their school's science club, if possible, for practical experience.

Suggested High School Subjects

- Algebra
- Biology
- Calculus
- Earth or Life or Physical Science
- Geometry
- Physics
- Trigonometry

Related Career Pathways/Majors

Science, Technology, Engineering & Mathematics Career Cluster

- Engineering & Technology Pathway

Postsecondary

Entry-level jobs for health and safety engineers typically require a bachelor's degree in an engineering discipline, such as chemical, mechanical, industrial, or systems engineering, or in environmental health and safety. Bachelor's degree programs typically include classroom, laboratory, and field studies in applied engineering. Engineering students interested in becoming health and safety engineers also should take courses in occupational safety and health, industrial hygiene, ergonomics, or environmental

Transferable Skills and Abilities

Communication Skills

- Interpreting federal and state regulations and their intent so that they can propose proper designs for specific work environments
- Preparing and presenting training materials to workers
- Describing new regulations and procedures to a variety of audiences

Creativity

- Producing designs showing potential problems and remedies for them
- Dealing with situations that are unique to a project

Critical-thinking Skills

- Identifying hazards to humans and property in the workplace or in the home before those hazards cause material damage or become a health threat

Observational Skills

- Observing and learning how operations function so that they can identify risks to people and property
- Recommending systemic changes to minimize risks

Problem-solving Skills

- Designing solutions for entire organizational operations
- Taking into account processes from more than one system at the same time
- Anticipating a range of human reactions to the changes they recommend

safety. Employers may prefer to hire graduates of engineering programs accredited by a professional association such as the Accreditation Board for Engineering and Technology, Inc. (ABET).

Some colleges and universities offer cooperative-education programs that allow students to gain practical experience while completing their education. Students also may choose to participate in an internship.

A few colleges and universities offer five-year accelerated programs through which students graduate with both a bachelor's and a master's degree. A master's degree allows engineers to enter the occupation at a higher level, from which they can develop and implement safety systems.

Related College Majors

- Chemical Engineering
- Environmental Health & Safety
- Health Sciences
- Industrial Engineering
- Mechanical Engineering
- Systems Engineering

Adult Job Seekers

Adults interested in going into health and safety engineering will find that they typically need a bachelor's degree in an engineering discipline or in environmental health and safety. Many employers in this field also value practical experience, and cooperative programs or internships are consequently also very beneficial for job seekers.

Professional Certification and Licensure

Licensure is not required for entry-level positions as a health and safety engineer. A Professional Engineering (PE) license, which allows for higher levels of leadership and independence, can be acquired later in one's career.

Health and safety engineers can earn professional certifications, including the following:

- The Board of Certified Safety Professionals offers the Certified Safety Professional (CSP) certification, the Occupational Health and Safety Technologist (OHST) certification, and the new Associate Safety Professional (ASP) certification

- The American Board of Industrial Hygiene awards the Certified Industrial Hygienist (CIH) certification
- The American Society of Safety Professionals offers a Certificate in Safety Management (CSM)
- The International Council on Systems Engineering offers a program leading to designation as a Certified Systems Engineering Professional (CSEP)

Additional Requirements

Employers may offer new employees training in the type of specialized editing software they use. Most editors eventually specialize in one type of software, but beginners should be familiar with as many types as possible.

EARNINGS AND ADVANCEMENT

The median annual wage for health and safety engineers was $94,240 in 2020. The lowest 10 percent earned less than $55,390, and the highest 10 percent earned more than $144,800.

Agricultural engineers may receive benefits such as paid vacations, holidays, and sick days; life and health insurance; and retirement benefits, usually paid by the employer.

Most health and safety engineers work full time.

EMPLOYMENT AND OUTLOOK

Employment of health and safety engineers is projected to grow 6 percent from 2020 to 2030, about as fast as the average for all occupations.

Health and safety engineers are employed mainly in manufacturing, engineering and consulting firms, construction, and state and local government. As buildings, products, and processes continue to become more complex and new regulations are created, these engineers will be needed to reduce costs, save lives, and produce safe consumer products.

Related Occupations

- Construction & Building Inspector
- Fire Inspector
- Industrial Engineer

- Mining/Geological Engineer
- Occupational Health/Safety Specialist/Technician

MORE INFORMATION

Accreditation Board for Engineering and Technology, Inc. (ABET)
415 N. Charles Street
Baltimore, MD 21201
410.347.7700
comms@abet.org
www.abet.org

American Board of Industrial Hygiene (ABIH)
6005 W. St. Joseph Highway, Suite 300
Lansing, MI 48917
517.321.2638
www.abih.org

American Industrial Hygiene Association (AIHA)
3141 Fairview Park Drive, Suite 777
Falls Church, VA 22042
703.849.8888
infonet@aiha.org
www.aiha.org

American Society of Safety Professionals (ASSP)
520 N. Northwest Highway
Park Ridge, IL 60068
847.699.2929
customerservice@assp.org
www.assp.org

Board of Certified Safety Professions (BCSP)
8645 Guion Road
Indianapolis, IN 46268
317.593.4800
www.bcsp.org

Daniel Vest

Conversation With...
SUBENA COLLIGAN, CIH, CSP

Industrial Hygienist
Principal, S. Colligan Coaching, Atlanta, GA
In the field 12 years

What was your individual career path in terms of education/training, entry-level job, or other significant opportunity?

My career in occupational and environmental health and safety (OEHS) happened by accident. I joined the military and wanted to go into the medical field, with the eventual goal of being a physician's assistant. My mom agreed that career would fit because I have horrible handwriting! When it was time for me to pick my job choices, the military career counselor suggested that I try a career in OEHS, also known as industrial hygiene. Since then, I have become a Certified Industrial Hygienist supporting a manufacturing company that makes planes, in research and development for a company that is changing how plastic enters the environment, and now I own my own consulting business. I have gotten the opportunity to change disciplines every two to three years and stay extremely excited to be an industrial hygienist.

What are the most important skills and/or qualities for someone in your profession?

The great advantage to the field is that there are so many different concentrations that skills and qualities in the profession vary. One consistent skill is a love for science and communication, written or oral. There are options to work directly with people on a daily basis, which requires empathy. There are jobs that are constantly changing, and others that work best for people who like structured challenges.

What do you wish you had known going into this profession?

I wish I knew the field existed before I entered. I would have committed in high school to be an industrial hygienist. As an entry level OEHS professional, I wish I knew just how vast the options were, and the many options I had for a career path.

Are there many job opportunities in your profession? In what specific areas?

Yes! There are options across the world (and in space). We can work in academia; at manufacturers supporting incredible missions like space travel and sustainable energy; the government to develop and enforce the regulations to protect people and the environment; and in the Peace Corps to support nations around the world. Any industry that employs people or impacts the environment needs individuals knowledgeable about protecting their workers and the environment.

How do you see your profession changing in the next five years, how will technology shift, and what skills will be required?

The career field will continue to grow. There is an increased emphasis on environment, sustainability, and governance around the world, and COVID-19 prioritized protecting people, with employees even demanding greater protections. Universities are starting to teach the principles of the field as required coursework for various fields of study outside of industrial hygiene, which further demonstrates just how important occupational and environmental safety and health is to the world.

What do you enjoy most about your job? What do you enjoy least about your job?

I enjoy the challenge of my job. I get to use critical thinking, research, personal interactions, and leadership, all while improving the livelihood of others. The least enjoyable part of my job is responding to someone who was injured on the job, or needing to investigate an unfavorable impact to the environment. It is everything my colleagues and I work to eliminate, but it still happens. I also know that our work lessens the overall number of environmental impacts and injuries, so I take a lot of pride in that.

Can you suggest a valuable "try this" for students considering a career in your profession?

If you want to get involved or learn more about the field, try this:

1. Find a university that offers industrial hygiene, environmental, or occupational safety as a major

2. Reach out to the department head and ask if you can shadow a hands-on activity (such as a group field trip)

Or

1. Find a local section of the American Industrial Hygiene Association (AIHA) (www.aiha.org/get-involved/local-sections/local-section-websites) or the American Society of Safety Professionals (ASSP) (www.assp.org/membership/communities/local-chapters) and a representative to come to your school to do an activity, or share their career.

As a profession, we love to give back, volunteer, and share our passion with everyone.

Conversation With...
VISHAL D. NATHU

Senior Manager
Corporate WHS Program Management
Amazon (AMZL—The Last Mile)
In the field, 4 years

What was your individual career path in terms of education/training, entry-level job, or other significant opportunity?

I hold both Bachelor's and Master's degrees in public health from the University of Kentucky and Eastern Kentucky University, respectively. My Master's degree was technically in two concentrations: industrial hygiene and environmental health. Currently, I am finishing my Ph.D. in the College of Medicine at the University of Cincinnati, focusing on biological aerosols—industrial hygiene.

My career has been incredibly unique and rewarding—I developed myself quickly at a very young age, and have grown in areas I never thought were imaginable. This is all because I started very young, and mastered time management skills (i.e. working full-time within my professional career while also being in graduate school full-time during my Master's and Doctoral degrees). Career-wise, I started in vehicle manufacturing at Toyota Motor North America, then practiced aerospace for a couple of years at Barnes Group. Now, I have finally found my home at Amazon, where I can dream big, make history, and deliver smiles to our customers worldwide.

What are the most important skills and/or qualities for someone in your Profession?

Occupational and Environmental Health and Safety (OEHS), also known as industrial hygiene, is an attractive but rigorous career path and choice. This is because there are potentially hazardous exposures and lives at stake within different workplaces. Modern-day OEHS requires in-depth comprehension of all natural sciences (chemistry, biology, physics, toxicology, etc.), and mathematics & statistics. Other knowledge such as epidemiology is also good to have in your pocket. These are skills that you must grasp in education, and you must then sharpen them in the field with your primary customer, the worker. Take note of how I said "modern-day"—you must also be proficient at communication (both verbal and written), since this career is exceptionally technical and precise. This career involves working with other people non-stop. You must be open to new ideas and be a good listener to help the worker achieve a safe and healthy working environment.

What do you wish you had known going into this Profession?

I wish I was at the very least briefed on the future of our profession. It has expanded so fast and is continuing to push boundaries every day. These are all good things, and I love it, but I wish I learned innovation while in school to get ahead of the curve. Luckily, due to the organizations I have been a part of, I was able to adapt quickly, practice new skills, and think at the next level (hence the Ph.D.); to shape my future and the future of my profession—and the association that represents us, the American Industrial Hygiene Association (AIHA). Every day is new for me—I never know what will happen or what my team will get into—however, one thing is for sure, I have never had one single dull day in my career.

Are there many job opportunities in your Profession? In what specific areas?

There are more job opportunities in my profession than people to fill the roles! Simply conduct a Google search and see for yourself: for instance, try "Workplace health and safety roles at Amazon," and see the many opportunities that appear. You can work anywhere, including retail, manufacturing, consulting, healthcare, governance, etc. The world is vast in the field of OEHS and industrial hygiene, and the opportunities are endless.

How do you see your Profession changing in the next five years, how will technology shift, and what skills will be required?

The field of OEHS/industrial hygiene is changing literally every day. I constantly read of new knowledge, best practices, new wearable technology, and new sensors that aid in workplace health and safety. New, up-and-coming occupational health professionals will have to excel in the classroom and must be comfortable with leading teams, being the face of an organization, and having outstanding people skills.

What do you enjoy most about your job? What do you enjoy least about your job?

I enjoy hearing experiences from our associates on how the workplace health and safety teams have shaped their process paths. I am a people-person, and love to engage with others, deliver presentations, speeches, and guide organizations in the right step. I also enjoy working with my team directly and helping one another grow. Perhaps the best part of my career is always looking forward to waking up and starting my day, rather than dreading it like some of my peers in other professions. Unfortunately, the worst part about this profession is seeing when occupational injuries and illness occur. I am a firm believer that all workplace injuries and illnesses are preventable, and we work hard every day to train people correctly to achieve that goal—"Striving to be Earth's Best Employer." We are innovators, deep thinkers, and always insist on the highest standard.

Can you suggest a valuable "try this" for students considering a career in your Profession?

Check out AIHA's page on our profession (www.OEHSCareers.org) and watch our videos (www.aiha.org/ih-careers/discover-industrial-hygiene/healthier-workplaces-a-healthier-world). Also, a simple Google or YouTube search of "industrial hygiene" or "occupational health" will give you an excellent taste of how much fun you can have in our profession. Lastly, just like most careers, you can always reach out to a professional and seek permission to shadow them, or even interview them.

Industrial Engineer

Snapshot

Career Cluster(s): Architecture & Construction; Manufacturing; Science, Technology, Engineering & Mathematics; Transportation, Distribution & Logistics

Interests: Critical Thinking; Routine Work; Analyzing Data; Decision-making; Leading Others

Earnings (Yearly Average): $88,950

Employment & Outlook: Faster Than Average Growth Expected

OVERVIEW

Sphere of Work

Industrial engineers find ways to eliminate wastefulness in production processes. They devise efficient systems that integrate workers, machines, materials, information, and energy to make a product or provide a service.

Work Environment

Depending on their tasks, industrial engineers work either in offices or in the settings they are trying to improve. For example, when observing problems, they may watch workers assembling parts in a factory. When solving problems, industrial engineers may be in an office at a computer where they analyze data that they or others have collected.

Photo via iStock/skynesher. [Used under license.]

Occupation Interest

Industrial engineers use their problem-solving skills to make the production processes of the products and services we use daily as efficient as possible. In general, they have an interest in using data, analytics, and creative-thinking in their work to provide solutions, test new concepts, and make improvements to preexisting systems. In addition, industrial engineers find fulfillment through working with numbers by conducting financial analyses and finding ways to save their clients money by eliminating unnecessary costs. They may be interested in working in more than one particular industry; industrial engineers are utilized in many different settings, including manufacturing, business, government, and more. These individuals are team players and enjoy working collaboratively with other engineers to brainstorm solutions to production problems.

Profile

Profile
Working Conditions: Inside Work
Physical Strength: Light Work
Education Needs: Bachelor's Degree
Licensure/Certification: Optional
Opportunities for Experience: Internship, Apprenticeship, On-the-Job Training
Interest Score: ICE

A Day in the Life—Duties and Responsibilities

Industrial engineers are the people behind production processes running smoothly. They use their knowledge of engineering and analytical minds to find solutions to both existing and potential problems. The systems they create or improve for making products and services perfectly integrate workers, machines, materials, information, and energy. They help to ensure that supply can meet demand by creating a capable system.

Industrial engineers apply their skills to many different situations, from manufacturing to healthcare systems to business administration.

Some industrial engineers, called manufacturing engineers, focus entirely on the automated aspects of manufacturing processes. They design manufacturing systems to optimize the use of computer networks, robots, and materials.

Industrial engineers focus on how to get the work done most efficiently, balancing many factors, such as time, number of workers needed, available technology, ac-

Duties and Responsibilities

- Reviewing production schedules, engineering specifications, process flows, and other information
- Understanding methods that are applied and activities that take place in manufacturing and services
- Figuring out how to manufacture parts or products, or deliver services, with maximum efficiency
- Developing management control systems to make financial planning and cost analysis more efficient
- Enacting quality control procedures to resolve production problems or minimize costs
- Designing control systems to coordinate activities and production planning
- Ensuring that products meet quality standards
- Conferring with clients about product specifications, vendors about purchases, management personnel about manufacturing capabilities, and staff about the status of projects

tions workers need to take, achieving the end product with no errors, workers' safety, environmental concerns, and cost.

The versatility of industrial engineers allows them to engage in activities that are useful to a variety of businesses, governments, and nonprofits. For example, industrial engineers engage in supply-chain management to help businesses minimize inventory costs; conduct quality assurance activities to help businesses keep their customer bases satisfied; and work in the growing field of project management as industries across the economy seek to control costs and maximize efficiencies.

WORK ENVIRONMENT

Immediate Physical Environment

Industrial engineers may find themselves working in both office settings and directly in the setting they are working to improve. If an industrial engineer mainly works in manufacturing, they may spend more time analyzing and observing a product's assembly in factories. However, an industrial engineer in the healthcare industry may spend much of their time observing hospitals, care clinics, or medical research centers. When finding solutions to problems, an industrial engineer is most likely to be found in an office space. Here, they can have access to computers for analyzing collected data and brainstorming ideas.

Industrial engineers must work well on teams because they need help from others to collect information about problems and to implement solutions.

Industrial engineers may need to travel to observe processes and make assessments in various work settings.

Human Environment

Industrial engineering is a very collaborative career path. It requires daily communication with other engineers, clients, staff, and a great variety of other workers and people from various backgrounds. These individuals are comfortable operating as part of a team to brainstorm ways to improve existing production processes and create new ones. Being able to communicate ideas effectively and in a way that can be easily understood by people with little or no engineering and scientific background is extremely important. In addition, they have very good listening skills and note the concerns of others, such as clients or customers. This is imperative to making sure that they can implement the feedback they are given in an effective way.

Technological Environment

Industrial engineers rely on a variety of technology and tools to complete their work. Some of the computer software they use includes computer-aided design (CAD) software, analytical and scientific software, industrial control software, supply-chain management software, and Office suite software. They also have a proficient understanding of the types of hardware and machinery that are used for creating prototypes and finished products.

EDUCATION, TRAINING, AND ADVANCEMENT

High School/Secondary

High school students interested in industrial engineering should take classes in mathematics, such as algebra, trigonometry, and calculus; computer science; and sciences, such as chemistry and physics. Typically, an industrial engineer will use these subjects to carry out most or all of their projects. Taking English and public speaking courses are very valuable for sharpening speaking and writing skills. Industrial engineers may use many different types of software to design and complete projects; courses in computer science or computer programming may give a student a head start on understanding these complex systems.

Suggested High School Subjects

- Algebra
- Biology
- Calculus
- Chemistry
- Computer Programming
- Earth/Life/Physical Science
- Economics
- English
- Entrepreneurship
- Geometry
- History
- Math Analysis
- Physics
- Psychology

- Statistics
- Trigonometry
- World Geography

Related Career Pathways/Majors

Architecture & Construction Career Cluster

- Design/Pre-Construction Career Pathway

Manufacturing Career Cluster

- Manufacturing Production Process Development Career Pathway

Science, Technology, Engineering & Mathematics Career Cluster

- Engineering & Technology Career Pathway

Transportation, Distribution & Logistics Career Cluster

- Warehousing and Distribution Center Operations Career Pathway

Postsecondary

Industrial engineers typically need a bachelor's degree in industrial engineering or industrial engineering technologies. However, many industrial engineers have degrees in mechanical engineering, electrical engineering, manufacturing engineering, or general engineering.

Bachelor's degree programs include lectures in classrooms and practice in laboratories. Courses include statistics, production systems planning, and manufacturing systems design, among others. Many colleges and universities offer cooperative education programs in which students gain practical experience while completing their education.

Transferable Skills and Abilities

Creativity

- Using creativity and ingenuity to design new production processes in many kinds of settings
- Reducing the use of material resources, time, or labor while accomplishing design goals

Critical-thinking Skills

- Creating new systems to solve problems related to waste and inefficiency
- Using logic and reasoning to identify strengths and weaknesses of alternative solutions, conclusions, or approaches to problems

Listening Skills

- Operating in teams
- Soliciting feedback from customers, vendors, and productions staff
- Listening to customers and clients in order to fully grasp ideas and problems

Math Skills

- Using the principles of calculus, trigonometry, and other advanced topics in mathematics for analysis, design, and troubleshooting in their work

Problem-solving Skills

- Dealing with several issues at once, from workers' safety to quality assurance

Speaking Skills

- Explaining instructions to production staff or technicians before they can make written instructions available
- Explaining concepts clearly and quickly to prevent costly mistakes and loss of time

Writing Skills

- Preparing documentation for other engineers or scientists, or for future reference
- Explaining thinking clearly so that others can understand the information

Several colleges and universities offer five-year degree programs in industrial engineering that lead to a bachelor's and master's degree upon completion, and several more offer similar programs in mechanical engineering. A graduate degree allows an engineer to work as a professor at a college or university or to engage in research and development (R&D). Some five-year or even six-year cooperative education plans combine classroom study with practical work, permitting students to gain experience and to finance part of their education.

Programs in industrial engineering are accredited by the Accreditation Board for Engineering and Technology, Inc. (ABET).

Related College Majors

- Chemical Engineering
- Civil Engineering
- Computer Engineering
- Electrical Engineering
- Industrial Engineering Technology
- Mechanical Engineering

Adult Job Seekers

Adults looking to work in this field should have experience in a similar or related occupation, as well as a bachelor's degree in industrial engineering or a subject related to it. Because industrial engineering is constantly advancing, former industrial engineers seeking to reenter the field can take local college courses to freshen up their knowledge on the subject.

Adult job seekers who are missing a related degree or applicable experience must enroll in relevant continuing education courses or in a postsecondary accredited college or university. A bachelor's degree in industrial engineering or a similar subject is necessary to work in this field.

Previously qualified industrial engineers may find work through a variety of sources, such as professional associations. Networking through engineering societies may provide both new and seasoned industrial engineers with broad opportunities.

Professional Certification and Licensure

Licensure is not required for entry-level positions as an industrial engineer. A Professional Engineering (PE) license, which allows for higher levels of leadership and independence, can be acquired later in one's career. Licensed engineers are called professional engineers (PEs). A PE can oversee the work of other engineers, sign off on projects, and provide services directly to the public. State licensure generally requires a degree from an ABET-accredited engineering program; a passing score on the Fundamentals of Engineering (FE) exam; relevant work experience, typically at least four years; and a passing score on the Professional Engineering (PE) exam.

The initial FE exam can be taken after one earns a bachelor's degree. Engineers who pass this exam are commonly called engineers in training (EITs) or engineer interns (EIs). After meeting work experience requirements, EITs and EIs can take the second exam, called the Principles and Practice of Engineering.

Each state issues its own licenses. Most states recognize licensure from other states, as long as the licensing state's requirements meet or exceed their own licensure requirements. Several states require continuing education for engineers to keep their licenses.

The Society of Manufacturing Engineers offers certification, which requires a minimum of eight years of a combination of education related to manufacturing and at least four years of work experience.

Additional Requirements

Industrial engineering is continuously evolving as new technology and research are produced each year. These professionals, in addition to having an advanced understanding of math and science, must be capable of evolving with the profession to ensure that they are always providing the most up-to-date and efficient solutions and systems. In short, they must be willing to continue learning throughout their career.

EARNINGS AND ADVANCEMENT

Earnings of industrial engineers depend on their education and experience, employer, licensure, and the types of projects on which they work. Median annual earnings of industrial engineers were $88,950 in 2020. The lowest 10 percent earned less than $57,950, and the highest 10 percent earned more than $136,930.

Industrial engineers who are just starting out usually work under the supervision of experienced engineers. In large companies, new engineers also may receive formal training in classes or seminars. As beginning engineers gain knowledge

Famous First

Henry Ford (1863-1947), founder of the automotive company bearing his name, was an industrial engineer perhaps best remembered for installing the first moving assembly line for the mass production of an entire automobile. His ground-breaking innovation cut the time it took to build a car from more than 12 hours to 2 hours and 30 minutes. Source: ppi2pass.com

Photo via iStock/Lalocracio. [Used under license.]

and experience, they move on to more difficult projects with greater independence to develop designs, solve problems, and make decisions.

Eventually, industrial engineers may advance to become technical specialists, such as quality engineers or facility planners. In that role, they supervise a team of engineers and technicians. Earning a master's degree facilitates such specialization and thus advancement.

Many industrial engineers move into management positions because the work they do is closely related to the work of managers.

EMPLOYMENT AND OUTLOOK

There were approximately 292,000 industrial engineers employed nationally in 2020. Employment of industrial engineers is expected to grow faster than the average for all occupations through the year 2030, at a rate of 14 percent.

This occupation is versatile both in the nature of the work it does and in the industries in which its expertise can be put to use.

Because they are not as specialized as other engineers, industrial engineers are employed in a wide range of industries, including major manufacturing industries, consulting and engineering services, and R&D firms. This versatility arises from the fact that these engineers focus on reducing internal costs, making their work valuable for many industries. For example, their work is important for manufacturing industries that are considering relocating from overseas to domestic sites.

In addition, growth in healthcare and changes in how healthcare is delivered will create demand for industrial engineers in firms in professional, scientific, and consulting services.

Related Occupations

- Architectural/Engineering Manager
- Cost Estimator
- Health & Safety Engineer
- Industrial Engineering Technologist/Technician
- Industrial Production Manager
- Logistician
- Management Analyst
- Occupational Health/Safety Specialist/Technician
- Quality Control Inspector

MORE INFORMATION

Accreditation Board for Engineering and Technology, Inc. (ABET)
415 N. Charles Street
Baltimore, MD 21201
410.347.7700
comms@abet.org
www.abet.org

American Society for Engineering Education (ASEE)
1818 N Street, NW, Suite 600
Washington, DC 20036
202.331.3500
www.asee.org

Institute of Industrial & Systems Engineers (IISE)
3577 Parkway Lane, Suite 200
Norcross, GA 30092
800.494.0460
cs@iise.org
www.iise.org

National Council of Examiners for Engineering and Surveying (NCEES)
200 Verdae Boulevard
Greenville, SC 29607
800.250.3196
www.ncees.org

National Society of Professional Engineers (NSPE)
1420 King Street
Alexandria, VA 22314
888.285.6773
www.nspe.org

Technology Student Association (TSA)
1904 Association Drive
Reston, VA 20191-1540
703.860.9000
general@tsaweb.org
www.tsaweb.org

Emily Porter

Conversation With...
ERIN GATELY

Director of Alliance Development and
Manager of Epeat Conformity Assurance
Green Electronics Council, Portland, OR
Industrial Engineer, 24 years

What was your individual career path in terms of education/training, entry-level job, or other significant opportunity?

I have an industrial engineering bachelor's from the University of Miami. I started working for Hewlett-Packard Company (HP) in 1991. Eventually, I moved into the marketing department, but it was for an engineering job that happened to be in that department. Industrial engineering deals with increasing efficiency and logistics.

I was working in marketing for a giant company trying to sell more things to more people, and that was very much at odds with my personal values. I had been taking classes in living simply with the Northwest Earth Institute. I took a leave of absence from HP and thought about how I could align my job with my values. I approached one of the institute's founders and said, "Hey, I took all these classes with you and now you need to help me figure out what to do for a living." When I told her I worked for Hewlett-Packard, she said, "Well you need to stay there because we need more people like you in big companies."

So I returned to HP and got a job as an Environmental Product Steward. It was my dream job. It was using my engineering skills, it was trying to build more environmentally friendly products —but I was terrible at it. I was like an evangelist. I'm sure the people I was in meetings with were rolling their eyes. I realized I didn't have the language to speak to business people, so I got an M.B.A. in 2004. My language totally changed. It wasn't "the right thing to do," it was "return on investment." Eventually I got a promotion and didn't like that job, so I left.

Now I'm like an investigator. I work with companies to make sure they're telling the truth. It's technical, but it's not pure engineering. But I do have to understand the technical specifications, so I need that engineering background.

What are the most important skills and/or qualities for someone in your profession?

Being willing to see new ways of doing things. Being really curious about "Is there a different way to do this? An environmentally friendly way?" You also have to have a commitment to lifelong education. I didn't stop at an engineering degree. I recently got a certificate in green chemistry, just for fun. Whatever engineering field you're working in, stay current with the exciting things that are happening in that discipline.

What do you wish you had known going into this profession?

I think the most important thing is something a manager said to me, which is that I am the driver of my own career. If you don't believe that, you'll be pushed along with the currents. You really have to stick up for yourself, especially as a woman, in all aspects of your career, whether that's speaking up in a meeting or making sure you're getting paid what you're worth. I'm generalizing, but often women are too nice. Again I'm generalizing, but a man wouldn't hesitate to say, "Hey, I did a great job on that. Can I get a raise?"

Are there many job opportunities in your profession? In what specific areas?

A lot of industrial engineering jobs are in manufacturing, of which there's not a ton in the U.S. I think it would be easier to find a job in industrial design.

How do you see your profession changing in the next five years? What role will technology play in those changes, and what skills will be required?

The Director of Alliance Development part of my job is all about looking for new, greener technologies. Companies are taking used toner and turning it into oil to run the machines in their factories. Toshiba is figuring out how to de-ink their paper so they can re-use the paper. Green products are here to stay.

What do you enjoy most about your job? What do you enjoy least about your job?

The thing that I like most is that it's like a puzzle. I'm always trying to figure out the most efficient ways to do things. What I like least is when it becomes rote. I said before the Environmental Product Stewardship job was my dream job-and it really was for about seven years. Then I was just doing the same thing over and over again.

Can you suggest a valuable "try this" for students considering a career in your profession?

Take something you do routinely and break it down into its component parts and see if there's any way you can make it more efficient. It can be something as simple as making a sandwich. Do you go to the fridge and get the meat out and put it on the counter, then go back to the fridge and get the mustard? If you enjoy figuring out things like this, you'll enjoy industrial engineering.

This interview was originally published in 2015.

Industrial Engineering Technologist/Technician

Snapshot

Career Cluster(s): Manufacturing; Science, Technology, Engineering & Mathematics; Transportation, Distribution & Logistics

Interests: Analytical Thinking; Problem-solving; Research; Routine Work; Analyzing Data

Earnings (Yearly Average): $57,320

Employment & Outlook: Slower Than Average Growth Expected

OVERVIEW

Sphere of Work

Industrial engineering technologists and technicians help engineers solve problems affecting manufacturing layout or production. They prepare machinery and equipment plans, design workflows, conduct statistical production studies, and analyze production costs.

Work Environment

Industrial engineers usually ask industrial engineering technologists and technicians to help carry out studies and draw conclusions. Consequently, these technologists and technicians typically work at the location where products are manufactured or services are provided.

Photo via iStock/NanoStockk. [Used under license.]

Occupation Interest

Industrial engineering technologists and technicians value using the principles of mathematics and science to solve a variety of production problems. These professionals help head engineers take their ideas and apply them to a production process by preparing plans and operations, analyzing the efficiency of the process, and finding ways to minimize costs. Industrial engineering technologists and technicians typically have a keen attention to detail and a knack for collecting and analyzing data to find areas that could use improvements.

Profile

Working Conditions: Inside
Physical Strength: Light Work; Medium Work
Education Needs: Postsecondary Certificate; Associate's Degree
Licensure/Certification: Not Required
Opportunities for Experience: Internship, On-the-Job Training
Interest Score: IRC

A Day in the Life—Duties and Responsibilities

Industrial engineering technologists and technicians help carry out studies and make observations to assist industrial engineers. They interpret schematic diagrams and formulas and apply them to the production of goods and services. Industrial engineering technicians are the workers that prepare the production process beforehand and analyze it afterwards. They are the team members that ensure everything in an operation runs as smoothly and efficiently as possible, and with the least amount of cost.

Duties and Responsibilities

- Suggesting revisions to operation methods, material handling, or equipment layout
- Interpreting engineering drawings, schematic diagrams, and formulas
- Conferring with management or engineering staff on quality and reliability standards
- Helping plan work assignments, considering factors such as machine capabilities and production schedules
- Preparing charts, diagrams, and other graphs to illustrate workflow, routing, floor layouts, how materials are handled, and how machines are used
- Collecting data to assist in process improvement activities

Industrial engineering technologists and technicians study the time and steps workers take to do a task (time and motion studies). To set reasonable production rates, they analyze operations such as maintenance, production, and service.

The work of industrial engineering technologists and technicians is versatile and applicable to a variety of projects. For example, in supply-chain management, they help businesses minimize inventory costs; in quality assurance, they help with customer satisfaction; and in project management, they help to control costs and maximize efficiencies.

Industrial engineering technologists and technicians generally work on teams under the supervision of industrial engineers.

OCCUPATION SPECIALTIES

Manufacturing Engineering Technologist/Technician
Manufacturing engineering technologists and technicians work to raise production quality and profitability. By planning, testing, and custom-making industrial products, they help engineers improve manufacturing processes and output. They may assess prototypes, analyze machinery performance, or try new production methods.

WORK ENVIRONMENT

Immediate Physical Environment

Industrial engineering technologists and technicians conduct most of their work in industry settings, typically where a product is manufactured or a service is provided. Here, they can closely analyze the production process in real time, conducting studies and extracting data and results in person.

Because they work in close contact with heavy machinery and equipment, industrial engineering technologists and technicians may be required to wear personal protective equipment while performing their daily tasks.

Human Environment

The projects that industrial engineering technologists and technicians work on require communication with a diverse team of professionals, such as engineers, manufacturing workers, and other technologists and technicians. They are capable of clearly communicating their findings to supervisors, articulating exactly which problems arise and what data they've collected. Industrial engineering technicians work collaboratively with others to ensure a production process continues running proficiently.

Technological Environment

Industrial engineering technologists and technicians utilize many different types of computer software and technology on a daily basis. The software they use includes computer-aided design (CAD), computer-aided manufacturing (CAM), analytical and scientific, industrial control, and Office suite software. They also have a proficient understanding of the types of hardware and machinery that are used during the production process. Knowing how these tools and equipment operate are necessary for being able to identify what is causing problems and ways in which it can be fixed.

EDUCATION, TRAINING, AND ADVANCEMENT

High School/Secondary

High school students interested in becoming industrial engineering technologists and technicians should take courses in math, science, and drafting, where available. These subjects are essential to the role of an industrial engineering technician and are utilized daily. Taking English and public speaking courses are very valuable for sharpening communication and interpersonal skills. Because industrial engineering technologists and technicians use many different types of software, courses in computer science or computer programming may give a student an introductory understanding these complex systems.

Suggested High School Subjects

- Algebra
- Biology
- Calculus
- Chemistry
- Civics
- Computer Programming
- Earth or Life or Physical Science
- Economics
- English
- Entrepreneurship
- Geometry
- History
- Math Analysis
- Physics
- Psychology
- Statistics
- Trigonometry
- World Geography

Transferable Skills and Abilities

Analytical Skills

- Assessing changes in conditions, operations, and the environment to help industrial engineers figure out how systems should work

Communication Skills

- Listening carefully to instructions from engineers
- Clearly articulating problems to their supervisors

Critical-thinking Skills

- Identifying and correcting weaknesses to help industrial engineers solve problems

Detail-oriented

- Recording precisely what they measure and observe

Math Skills

- Using mathematics and statistical techniques to analyze data collected from studies

Observational Skills

- Watching the performance of people or organizations so that they can suggest improvements

Related Career Pathways/Majors

Manufacturing Career Cluster

- Manufacturing Production Process Development Career Pathway

Science, Technology, Engineering & Mathematics Career Cluster

- Engineering & Technology Career Pathway

Transportation, Distribution & Logistics Career Cluster

- Warehousing & Distribution Center Operations Career Pathway

Postsecondary

Postsecondary programs in industrial engineering technology are offered at vocational-technical schools, technical institutes, and community colleges. Vocational-technical schools typically award a certificate. Community colleges programs usually lead to associate's degrees.

Employers may prefer to hire candidates who have completed an engineering or engineering technology program accredited by the Accreditation Board for Engineering and Technology, Inc. (ABET).

Industrial engineering technologists and technicians may choose to major in applied science, industrial technology, or industrial engineering technology. These programs may include instruction in CAD/CAM software.

Related College Majors

- Applied Science
- Chemical Engineering Technology
- Engineering Technology
- Industrial Engineering Technology
- Industrial Technology
- Manufacturing Engineering Technology

Adult Job Seekers

Adults who do not have prior experience in industrial engineering technology should enroll in either an accredited associate's degree program or a postgraduate certification program. Although an associate's degree may be preferred by some employers, it is not always required. Many technical and vocational schools offer a certificate in industrial engineering technology while community colleges typically offer an associate's degree, both of which are valid credentials when searching for work in this field. Previous experience in a similar or related field will benefit an adult seeking a career in this profession.

Professional Certification and Licensure

Although certification or licensure is not required to work as an industrial engineering technologist or technician, it may help technologists and technicians develop their careers.

Additional Requirements

Industrial engineering technologists and technicians, in addition to performing well in mathematics and science, use their knowledge and expertise to provide solutions to a variety of problems in production processes. They should be comfortable taking direction from others and performing tasks that are delegated to them. In addition, they need to be able to fully interpret the data they collect, reading in-between the lines to find what might be causing production delays or below-quality products and services.

EARNINGS AND ADVANCEMENT

Earnings of industrial engineering technologists and technicians depend on their education and experience, employer, licensure, specialty, and the types of projects on which they work. Median annual earnings of industrial engineering technologists and technicians were $57,320 in 2020. The lowest 10 percent earned less than $36,550, and the highest 10 percent earned more than $86,900.

Industrial engineering technologists and technicians may receive benefits such as paid vacations, holidays, and sick days; life and health insurance; and retirement benefits, usually paid by the employer.

Technicians who obtain appropriate education or certification may advance to become technologists.

EMPLOYMENT AND OUTLOOK

There were approximately 64,100 industrial engineering technologists and technicians employed nationally in 2020. Employment of industrial engineering technologists and technicians is expected to grow slower than the average for all occupations through the year 2030, at a rate of 4 percent.

An emphasis on cost control through increased efficiency, along with industrial engineering technologists and technicians' role in assisting with automation, is expected to sustain demand somewhat for these workers.

However, overall employment growth of industrial engineering technologists and technicians in manufacturing—the industry in which most of them work—is projected to be slow.

Related Occupations

- Cost Estimator
- Health & Safety Engineer
- Industrial Engineer
- Logistician
- Quality Control Inspector

MORE INFORMATION

Accreditation Board for Engineering and Technology, Inc. (ABET)
415 N. Charles Street
Baltimore, MD 21201
410.347.7700
comms@abet.org
www.abet.org

American Society for Engineering Education (ASEE)
1818 N Street, NW, Suite 600
Washington, DC 20036
202.331.3500
www.asee.org

Institute of Industrial & Systems Engineers (IISE)
3577 Parkway Lane, Suite 200
Norcross, GA 30092
800.494.0460
www.iise.org

Society of Manufacturing Engineers (SME)
1000 Town Center, Suite 1910
Southfield, MI 48075
313.425.3000
www.sme.org

Emily Porter

Marine Engineer/ Naval Architect

Snapshot

Career Cluster(s): Architecture & Construction; Science, Technology, Engineering & Mathematics; Transportation, Distribution & Logistics

Interests: Design; Mathematics; Problem-solving; Analyzing Data; Decision-making

Earnings (Yearly Average): $95,440

Employment & Outlook: Slower Than Average Growth Expected

OVERVIEW

Sphere of Work

Marine engineers and naval architects design, build, and maintain ships, from aircraft carriers to submarines and from sailboats to tankers. Marine engineers are also known as marine design engineers or marine mechanical engineers and are responsible for the internal systems of a ship, such as the propulsion, electrical, refrigeration, and steering systems. Naval architects are responsible for the ship design, including the form, structure, and stability of hulls.

Photo via iStock/shaunl. [Used under license.]

Work Environment

Marine engineers and naval architects typically work in offices, where they have access to computer software and other tools necessary for analyzing projects and designing solutions.

Sometimes, they must go to sea to test or maintain the ships that they have designed or built.

Profile

Working Conditions: Both Inside and Outside
Physical Strength: Light Work
Education Needs: Bachelor's Degree
Licensure/Certification: Required
Opportunities for Experience: Cooperative Education Programs; Internship; On-the-Job Training
Interest Score: IRA

Occupation Interest

Marine engineers and naval architects take the concepts of mathematics and science and apply them creatively to real-world projects. They use their knowledge of these subjects to design the systems, interior, and exterior of a variety of vessels, such as naval, cruise, or cargo ships and tankers, submarines, and boats for recreational use. Marine engineers and naval architects working in government may be responsible for designing the ships used by the military. Because marine engineers and naval architects have to ensure that the vessels they create are perfectly sound, someone with high attention to detail will excel in this field. In addition, these professionals are not discouraged by high-pressure situations; in fact, they may find that they work best in these environments.

Duties and Responsibilities

Marine Engineers

- Preparing system layouts and detailed drawings and schematics
- Inspecting marine equipment and machinery
- Drawing up work requests and job specifications
- Conducting environmental, operational, or performance tests on marine machinery and equipment
- Designing and overseeing the testing, installation, and repair of marine equipment
- Investigating and testing machinery and equipment to ensure compliance with standards
- Coordinating activities with regulatory bodies to ensure that repairs and alterations are done safely and at minimal cost
- Preparing technical reports for use by engineers, managers, or sales personnel
- Preparing cost estimates, contract specifications, and design and construction schedules
- Maintaining contact with contractors to make sure that the work is being done correctly, on schedule, and within budget

A Day in the Life—Duties and Responsibilities

Marine engineers and naval architects are responsible for the process of creating an assortment of ships, from their conception to their execution. Depending on the sector they work in, marine engineers and naval architects may design large vessels such as cruise ships and tankers, or smaller vessels such as sailboats and lifeboats. Marine engineers in particular design the internal systems of ships. They are responsible for ensuring that the operations on a ship run smoothly by planning the propulsion, electrical, refrigeration, and steering systems, in addition to many others. The interiors, exteriors, and overall stability of a ship are the creation of naval architects.

Marine engineers and naval architects apply knowledge from a range of engineering fields to the entire water vehicles' design and production processes. Marine engineers also design and maintain offshore oil rigs and may work on alternative energy projects, such as wind turbines located offshore and tidal power.

Marine engineers and naval architects who work for ship- and boatbuilding firms design large ships, such as passenger ships and cargo ships, as well as small craft, such as inflatable boats and rowboats. Those who work in the federal government may design or test the designs of ships or systems for the Army, Navy, or Coast Guard.

WORK ENVIRONMENT

Duties and Responsibilities

Naval Architects

- Studying design proposals and specifications
- Establishing basic characteristics of a ship, such as its size, weight, and speed
- Developing sectional and waterline curves of the ship's hull to establish the center of gravity, the ideal hull form, and data on buoyancy and stability
- Designing entire ship hulls and superstructures, following safety and regulatory standards
- Designing the complete layout of ships' interiors, including spaces for machinery and auxiliary equipment, passenger compartments, cargo space, ladder wells, and elevators
- Designing the layout of boiler room equipment, heating and ventilation systems, refrigeration equipment, electrical distribution systems, safety systems, steering systems, and propulsion machinery
- Leading teams from a variety of specialties to oversee building and testing prototypes
- Evaluating how ships perform during trials, both in the dock and at sea
- Changing designs as needed to make sure that national and international standards are met

Immediate Physical Environment

Marine engineers and naval architects may conduct work in both office settings and on the ships they help design or build. When conceptualizing a ship's design, deliberating solutions to potential and impending problems, and analyzing the project they are carrying out, both marine engineers and naval architects typically work in an office where they have access to computer software and other essential tools. Occasionally, these professionals must attend to the ships they create to assist in the performance testing or maintenance of a vessel.

Marine engineers and naval architects who work on power-generation projects, such as offshore wind turbines and tidal power, work along the coast—both offshore and on land. They also sometimes work on oil rigs, where they oversee the repair or maintenance of systems that they may have designed.

Naval architects often lead teams to create feasible designs, and they must effectively use the skills that each person brings to the design process.

Human Environment

Marine engineers and naval architects communicate with many other people during the course of their work. To ensure plans, preparations, and construction are being adequately followed and executed, they coordinate with other engineers and architects, clients, construction crews, and when applicable, government officials. Not every person a marine engineer or naval architect communicates with will have the same experience in engineering or science, so they must be able to explain their designs and findings in a way that can be easily understood by people from all backgrounds. When they are leading projects, they must feel comfortable giving direction and feedback to the staff working alongside them.

Technological Environment

Marine engineers and naval architects are familiar with a broad variety of technology and tools used in completing designs and projects. They utilize multiple types of computer software to perform research, design ships, and analyze the data they collect. Such computer software ranges from computer-aided design (CAD) and analytical and scientific software to Office suite and project management software. If they visit a ship for maintenance and testing, they use tools for surveying the systems and structures. These could include hand and power tools as well as instruments for measuring the ship's stability and speed.

EDUCATION, TRAINING, AND ADVANCEMENT

High School/Secondary

High school students interested in becoming a marine engineer or naval architect should take classes in math, such as algebra, trigonometry, and calculus; and science, such as chemistry and physics. These subjects are used daily in the life of a marine engineer or naval architect. Taking English and public speaking courses are very valuable for sharpening communication and interpersonal skills. Because marine engineers and naval architects use many different types of software, courses in computer science or computer programming may give a student a jumpstart on understanding these complex systems. If available, drafting courses may be helpful for aspiring naval architects.

Suggested High School Subjects

- Algebra
- Biology
- Calculus
- Chemistry
- Computer Programming
- Earth or Life or Physical Science
- Economics
- English
- Entrepreneurship
- Geometry
- History
- Math Analysis
- Physics
- Psychology
- Statistics
- Trigonometry
- World Geography

Related Career Pathways/Majors

Architecture & Construction Career Cluster

- Design/Pre-Construction Career Pathway

Science, Technology, Engineering & Mathematics Career Cluster

- Engineering & Technology Career Pathway

Transportation, Distribution & Logistics Career Cluster

- Transportation Systems/Infrastructure Planning, Management, & Regulation Career Pathway

Transferable Skills and Abilities

Communication Skills

- Giving clear instructions
- Explaining complex concepts when leading projects

Ingenuity

- Using operations analysis to create a design to perform the ship's functions
- Employing critical-thinking skills to anticipate and correct any deficiencies before the ship is built or set to sea

Interpersonal Skills

- Meeting with clients to analyze their needs for ship systems
- Discussing progress with clients to keep redesign options open before the project is too far along

Math Skills

- Using the principles of calculus, trigonometry, and other advanced topics in math for analysis, design, and troubleshooting in their work

Problem-solving Skills

- Designing several systems that work well together in ships
- Solving problems for clients
- Drawing on knowledge and experience to make effective decisions

Postsecondary

College students interested in this occupation typically pursue a degree in an engineering field. Programs in marine engineering and naval architecture usually include courses in computer-aided design, fluid mechanics, ship hull strength, and mechanics of materials. Most programs also include time at sea, during which students gain hands-on experience on a vessel.

Some marine engineering and naval architecture programs are offered at state maritime academies. Students studying at the maritime academies spend time at sea, usually during the summer, to gain onboard operating experience. For more information about state maritime academies, visit the Maritime Administration of the U.S. Department of Transportation.

Employers may prefer to hire candidates who graduate from a program accredited by a professional association such as the Accreditation Board for Engineering and Technology, Inc. (ABET).

Related College Majors

- Chemical Engineering
- Civil Engineering
- Computer Engineering
- Electrical Engineering
- Marine Engineering
- Marine Engineering Technology
- Mechanical Engineering

Adult Job Seekers

Adults interested in marine engineering or naval architecture ideally have experience in a similar or related occupation, as well as a bachelor's degree in marine engineering or a subject related to it, such as mechanical engineering. Former marine engineers and naval architects seeking reentry into the field should take local college courses to freshen up their knowledge on the subject.

Adult job seekers without a related degree or applicable experience that are interested in becoming a marine engineer or naval architect must enroll in a postsecondary accredited college or university or in continuing education courses. A bachelor's degree in some branch of engineering is needed for anyone to work in this field.

Those that are qualified marine engineers and naval architects may find work through a variety of sources, including the government and professional associations. Networking through engineering societies may provide marine engineers and naval architects that have varying years of experience in the field with job opportunities and important connections.

Professional Certification and Licensure

Along with earning a bachelor's degree, students at states' maritime academies take exams for licensure from the U.S. Coast Guard.

Another type of engineering license is the Professional Engineering (PE) license, which allows for higher levels of leadership and independence and can be acquired later in one's career. Licensed engineers are called professional engineers (PEs). A PE can oversee the work of other engineers, sign off on projects, and provide services directly to the public. State licensure generally requires a degree from an ABET-accredited engineering program; a passing score on the Fundamentals of Engineering (FE) exam; relevant work experience, typically at least four years; and a passing score on the Professional Engineering (PE) exam.

The initial FE exam can be taken after earning a bachelor's degree. Engineers who pass this exam are commonly called engineers in training (EITs) or engineer interns (EIs). After meeting work experience requirements, EITs and EIs can take the second exam, called the Principles and Practice of Engineering (PE).

Additional Requirements

Marine engineers and naval architects have an advanced understanding of math and science and apply their knowledge creatively in designing systems and ships and providing solutions to various problems. Occasionally, marine engineers and naval architects will step into leadership positions to supervise projects and give direction. Marine engineering and naval architecture frequently evolve as new technology and research are produced every year. Professionals in this industry need to be capable of evolving with the profession in order to ensure that they are always providing the most up-to-date and efficient solutions and systems.

EARNINGS AND ADVANCEMENT

Earnings of marine engineers and naval architects depend on their education and experience, employer, licensure, specialty, and the types of projects on which they work. Median annual earnings of marine engineers and naval architects were $95,440 in 2020. The lowest 10 percent earned less than $62,610, and the highest 10 percent earned more than $148,580.

Marine engineers and naval architects may receive benefits such as paid vacations, holidays, and sick days; life and health insurance; and retirement benefits, usually paid by the employer.

Famous First

Ancient Greek mathematician and inventor Archimedes (287-211 BCE) is known to have developed several marine engineering systems, the principles of which still apply and are taught to students today. Modern marine engineering started in the 1700s, in the early days of the Industrial Revolution.
Source: planm8.io

Photo via iStock/FotoGablitz. [Used under license.]

Beginning marine engineers usually work under the supervision of experienced engineers. In larger companies, new engineers also may receive formal training in classrooms or seminars. As beginning engineers gain knowledge and experience, they move on to more difficult projects, on which they have greater independence to develop designs, solve problems, and make decisions.

Eventually, marine engineers may advance to become technical specialists or to supervise a team of engineers and technicians. Some may even become engineering managers or move into other managerial positions or sales work. In sales, an engineering background enables them to discuss technical aspects of certain kinds of engineering projects. Such knowledge is also useful in assisting clients in project planning, installation, and use.

EMPLOYMENT AND OUTLOOK

There were approximately 9,800 marine engineers and naval architects employed nationally in 2020. Employment of marine engineers and naval architects is expected to grow slower than the average for all occupations through the year 2030, at a rate of 4 percent.

Many jobs for these workers are in the ship- and boatbuilding industry, which supplies new ship orders for the military. Therefore, federal budgetary constraints may limit employment growth.

Technological developments have made offshore wind energy projects more feasible. As a result, more marine engineers and naval architects are expected to be needed to help plan and implement these projects.

Related Occupations

- Aerospace Engineer
- Electrical/Electronics Engineer
- Mechanical Engineer
- Petroleum Engineer

MORE INFORMATION

Accreditation Board for Engineering and Technology. Inc. (ABET)
415 N. Charles Street
Baltimore, MD 21201
410.347.7700
comms@abet.org
www.abet.org

American Society for Engineering Education (ASEE)
1818 N Street, NW, Suite 600
Washington, DC 20036
202.331.3500
www.asee.org

American Society of Naval Engineers (ASNE)
1423 Powhatan Street, Suite 1
Alexandria, VA 22314
703.836.6727
asnehq@navalengineers.org
www.navalengineers.org

Marine Engineers' Beneficial Association (MEBA)
444 N. Capitol Street, NW, Suite 800
Washington, DC 20001
202.638.5355
mebahq@mebaunion.org
www.mebaunion.org

Maritime Administration of the U.S. Department of Transportation
1200 New Jersey Avenue, SE
Washington, DC 20590
202.493.0029
careersafloat@dot.gov
www.maritime.dot.gov

Technology Student Association (TSA)
1904 Association Drive
Reston, VA 20191-1540
703.860.9000
general@tsaweb.org
www.tsaweb.org

Emily Porter

Conversation With...
MATTHEW TOMPKINS

Chief Operating Officer
TC Defense, Arlington, VA
Marine Engineer, 10 years

What was your individual career path in terms of education/training, entry-level job, or other significant opportunity?

I originally came from Indiana, where there's no ocean or maritime-related work, but when I decided to attend the U.S. Merchant Marine Academy in Kings Point, NY, on Long Island, I was more interested in the military aspect. I earned a B.S. in Marine Engineering and Shipyard Management. The academy requires a lot of hands-on, co-op experience, including spending 360 days at sea on a commercial merchant ship. I went to 16 different countries while still in college, which opened my eyes to different opportunities. There's also a four-week internship requirement, which showed me what the commercial marine industry has to offer. I interned with a small Indiana company that made small vessels for inland waterways.

My first job was working for BAE Systems as a contractor to Naval Sea Systems Command designing and building ships. The Navy would have a portfolio of systems, and $20 to $30 million worth of equipment that had to be installed in a specific time period. So I managed those efforts onsite, reviewed design drawings, walked the project, ensured proper installation and inspected for quality control.

I went on to work for some large defense companies and I've done project management. T.C. Defense is a family business I co-founded, and I'm now working on a Ph.D. in systems engineering from George Washington University. I'm trying to achieve more senior levels of technical consulting within the defense market. I'm also a Naval Reservist in the Military Sealift Command, which has a cadre of licensed marine engineers.

One of the great things about being a licensed marine engineer is that you get a handle on everything mechanical or electrical-related on a ship. A traditional maritime engineer is a mariner who operates a ship, but I haven't done that since I graduated.

What are the most important skills and/or qualities for someone in your profession?

Being capable of dealing with people at all levels, from skilled laborers to senior executives. You have to be adaptable, persistent, persuasive and patient.

What do you wish you had known going into this profession?

There's a big knowledge gap for me with certain detail design characteristics, and I wish I had known that gap was there. In addition, it was a huge shock to encounter bureaucracy-related waste, especially in government work. In addition, this is an age-dominated industry. There's not a lot of innovation and oftentimes I think there's too much value put on the number of years of experience a person has. That applies to both marine engineering and government. I've had to learn to work around that.

Are there many job opportunities in your profession? In what specific areas?

Yes. There's a shortage of licensed marine engineers in both the commercial and naval marine engineering markets. With a marine engineering license, you can operate commercial propulsion plants onboard ships or work shoreside in engineering fields, such as aerospace or defense engineering. Many of the systems and components you find on a ship are the same components found on other types of products, like airplanes.

How do you see your profession changing in the next five years, what role will technology play in those changes, and what skills will be required?

The biggest technological changes will be in how ships are fueled and powered. Commercial ships are seeking to lower fuel costs by using liquefied natural gas instead of diesel, and Navy ships are seeking to have more power for advanced warfighting capabilities.

What do you enjoy most about your job? What do you enjoy least about your job?

I get to choose whether I want to "geek out" on technical projects or stay on the project management side of things. There's a shortage of engineers and technical folks out there, so if somebody's willing to step up and do the technical work, organizations are willing to let you do that.

I least enjoy dealing with difficult people. When I went through engineering school, I was told that 90 percent of engineering is dealing with people and 10 percent is technical, and that's pretty much how it turned out.

Can you suggest a valuable "try this" for students considering a career in your profession?

Many of the nation's maritime universities have special days where they invite high school students to visit, get exposure to the curriculum and actually board a training ship. Take advantage of those opportunities!

This interview was originally published in 2015.

Materials Engineer

Snapshot

Career Cluster(s): Architecture & Construction; Science, Technology, Engineering & Mathematics

Interests: Mathematics; Science; Technology; Problem-Solving

Earnings (Yearly Average): $95,640

Employment & Outlook: As Fast as Average

OVERVIEW

Sphere of Work

Materials engineers develop, process, and test materials used to create a range of products, from computer chips and aircraft wings to golf clubs and biomedical devices. They study the properties and structures of metals, ceramics, plastics, composites, nanomaterials (extremely small substances), and other substances in order to create new materials that meet certain mechanical, electrical, and chemical requirements. They also help select materials for specific products and develop new ways to use existing materials.

Work Environment

Materials engineers often work in offices where they have access to computers and design equipment. Others work in factories or research and development (R&D) laboratories. Materials engineers may work in teams with scientists and engineers from other backgrounds.

Photo via iStock/EvgeniyShkolenko. [Used under license.]

Occupation Interest

For those interested in pursuing a career as a materials engineer, a strong knack or interest in science, mathematics, and engineering is important. Also, an individual must be able to troubleshoot complex problems and challenges, as they will be faced with instances during work that will require patience in determining the root of an issue when it comes to a project design and its testing.

Profile

Working Conditions: Mainly Indoors
Physical Strength: Light Work
Education Needs: Bachelor's Degree; Master's Degree; Doctoral Degree
Licensure/Certification: Required
Opportunities for Experience: Internship; Apprenticeship; Part-time Work
Interest Score: IRE

A Day in the Life—Duties and Responsibilities

The typical day for a materials engineer will vary according to specific subfields or specializations. While some engineers focus on materials design, others focus on the physical creation of new materials in laboratory settings. Some materials engineers divide their time between design and production activities.

In most cases, materials engineers work to produce materials that meet the specifications created by teams of individuals who work on other aspects of the design process; for instance, a materials engineer might work with a mechanical engineer to address requirements such as the density, flexibility, and mechanical strength of a specific material. Materials engineers also spend part of their time testing the characteristics of materials to determine how useful they will be for the manufacture of various objects. Materials engineers subject materials to a variety of highly specialized tests as part of the development process. In the aerospace field, for instance, materials used in the construction of aircraft wings must undergo wind testing to determine how well the materials hold up when subjected to conditions similar to those at high altitudes and speeds.

Duties and Responsibilities

- Planning and evaluating new projects, and consulting with other engineers and managers as necessary
- Preparing proposals and budgets, as well as analyzing labor costs, writing reports, and performing other managerial tasks
- Supervising the work of technologists, technicians, and other engineers and scientists
- Designing and directing the testing of processing procedures
- Monitoring how materials perform and evaluating how they deteriorate
- Determining causes of product failure and developing ways of overcoming such failure
- Evaluating technical specifications and economic factors relating to the design objectives of processes or products
- Evaluating the impact of materials processing on the environment

In many cases, engineers must prepare and conduct a variety of evaluations as they complete various aspects of their jobs. An engineer producing a new type of metal, for instance, must often produce reports on a variety of economic and budgetary factors, including labor and materials-sourcing costs and the estimated cost of producing materials given a variety of different design and component options. Engineers working in a managerial capacity must also supervise and evaluate the performance of subordinate staff.

OCCUPATION SPECIALTIES

Ceramic Engineer
Ceramic engineers develop ceramic materials and the processes for making them into useful products, from high-temperature rocket nozzles to glass for liquid-crystal display (LCD) flat-panel displays.

Composites Engineer
Composites engineers develop materials with special, engineered properties for applications in aircraft, automobiles, and related products.

Metallurgical Engineer
Metallurgical engineers specialize in metals, such as steel and aluminum, usually in alloyed form with additions of other elements to provide specific properties.

Fiber Artist
Fiber artists use fabric, yarn, or other natural and synthetic materials to weave, knit, crochet, or sew textile art. They may use a loom to weave fabric, needles to knit or crochet yarn, or a sewing machine to join pieces of fabric for quilts or other handicrafts.

Plastics Engineer
Plastics engineers develop and test new plastics, known as polymers, for new applications.

Semiconductor Processing Engineer
Furniture makers cut, sand, join, and finish wood and other materials to make handcrafted furnishings. For information about other workers who assemble wood furniture, see the profile on woodworkers.

WORK ENVIRONMENT

Immediate Physical Environment

Materials engineers often work in offices where they have access to computers and design equipment. Others work in factories or R&D laboratories. Materials engineers may work in teams with scientists and engineers from other backgrounds.

Human Environment

Materials engineers are oftentimes involved in a collaborative environment where they would work alongside fellow engineers, as well as technologists and technicians on a shared large-scale project. Depending on an individual's level of seniority, they might either be reporting to a supervisor, or be overlooking lower-level engineers.

Technological Environment

Engineering laboratories that focus on the designing and testing of projects often use highly specialized equipment specifically manufactured for certain tests or procedures. With the constant advancement of technology and equipment, engineers must frequently be open to engage in learning new software and equipment in their field. Also, the design aspect of a materials engineer's job will have them consistently working with computer-aided design (CAD) programs, as well as other programs that showcase the creation of three-dimensional (3D) models that will be worked on as part of the testing and development process.

EDUCATION, TRAINING, AND ADVANCEMENT

High School/Secondary

High school students interested in studying materials engineering should take classes in math, such as algebra, trigonometry, and calculus; science, such as biology, chemistry, and physics; and computer programming.

Suggested High School Subjects

- Algebra
- Applied Math
- Biology
- Calculus
- Chemistry

- Computer Science
- Earth Science
- English
- Geometry
- Mathematics
- Physics
- Statistics
- Trigonometry

Related Career Pathways/Majors

Architecture & Construction Career Cluster

- Design/Pre-Construction Pathway

Science, Technology, Engineering & Mathematics Career Cluster

- Engineering & Technology Pathway

Transferable Skills and Abilities

Analytical Skills

- Determining how materials will be used and how they must be structured to withstand different conditions

Math Skills

- Using the principles of calculus and other advanced topics in math for analysis, design, and troubleshooting in their work

Problem-solving Skills

- Understanding the relationship between materials' structures, their properties, how they are made, and how these factors affect the products they are used to make
- Figuring out why a product might have failed
- Designing a solution, then conducting tests to make sure that the product does not fail again
- Being able to identify root causes of an issue when many factors could be at fault

Speaking Skills

- Stating concepts and directions clearly when working with technicians, technologists, and other engineers
- Communicating engineering concepts to people who may not have an engineering background

Writing Skills

- Writing plans and reports clearly so that people without a materials engineering background can understand the concepts

Postsecondary

Some colleges and universities offer a five-year program leading to both a bachelor's and master's degree. A graduate degree allows an engineer to work as a postsecondary teacher or to do R&D.

Entry-level jobs for materials engineers typically require a bachelor's degree in engineering. Programs typically last four years and include classroom and laboratory work focusing on engineering principles.

Some colleges and universities offer a five-year program leading to both a bachelor's and master's degree. A graduate degree allows an engineer to work as a postsecondary teacher or to do R&D. Many colleges and universities offer internships and cooperative programs in partnership with industry employers. In these programs, students gain practical experience while completing their education.

Related College Majors

- Aerospace & Aeronautical Engineering
- Biomedical Engineering

Famous First

Swiss engineer George de Mestral (1907-1990) patented Velcro in 1955. It was inspired by nature, after he saw the way burdock burrs clung to other materials. Source: prezi.com

Photo via iStock/stocksnapper. [Used under license.]

- Chemical Engineering
- Civil Engineering
- Computer Engineering
- Electrical Engineering
- Environmental Engineering
- General Engineering
- Industrial Engineering
- Materials Engineering
- Materials Science
- Mechanical Engineering
- Metallurgical Engineering
- Metallurgy

Adult Job Seekers

Materials engineers typically need a bachelor's degree in materials science and engineering or in a related engineering field. Completing internships and cooperative-engineering programs while in school may be helpful for gaining hands-on experience.

Employers may prefer to hire graduates of engineering programs accredited by a professional association such as ABET (Accreditation Board for Engineering and Technology, Inc.). A degree from an accredited program is usually necessary to become a licensed professional engineer.

Professional Certification and Licensure

Licensure for materials engineers is not as common as it is for other engineering occupations, nor it is required for entry-level positions. A Professional Engineering (PE) license, which allows for higher levels of leadership and independence, can be acquired later in one's career. Licensed engineers are called professional engineers (PEs). A PE can oversee the work of other engineers, sign off on projects, and provide services directly to the public.

The initial Fundamentals of Engineering (FE) exam can be taken after earning a bachelor's degree. Engineers who pass this exam are commonly called engineers in training (EITs) or engineer interns (EIs). After meeting work experience requirements, EITs and EIs can take the second exam, called the Principles and Practice of Engineering (PE). Each state issues its own licenses. Most states recognize licensure from other states, as long as the licensing state's requirements meet or exceed their own licensure requirements. Several states require continuing education for engineers to keep their licenses.

Additional Requirements

Interpersonal communication skills are a necessary since many material engineers often work within a bigger team of fellow engineers, scientists, technicians, and technologists on large-scale projects. Additionally, an individual must have patience and ability to problem solve when issues come up in their designs and projects. Many times, the testing period is trial and error, so an individual has to be detail-oriented and be willing to study their mistakes in order to learn what went wrong and how to improve.

EARNINGS AND ADVANCEMENT

Median annual earnings of craft and fine artists were $95,640 in 2020. The lowest 10 percent earned less than $57,870, and the highest 10 percent earned more than $154,340.

There are many factors that play into how much an individual earns. This ranges from an individual's education, experience, and skill set. Furthermore, it can depend on the employer and where the employer is located, whether it is in a larger and more metropolitan city or a smaller area.

Junior materials engineers usually work under the supervision of experienced engineers. In large companies, new engineers may receive formal training in classrooms or seminars. As engineers gain knowledge and experience, they move on to more difficult projects where they have greater independence to develop designs, solve problems, and make decisions.

Eventually, materials engineers may advance to become technical specialists or to supervise a team of engineers and technicians. Many become engineering managers or move into other managerial positions or sales work. An engineering background is useful in sales because it enables sales engineers to discuss a product's technical aspects and assist in product planning, installation, and use.

Most materials engineers work full time. Some materials engineers work more than 40 hours per week. As full-time workers, materials engineers may receive paid holidays, vacations, and sick days. As well as retirement benefits, and insurance coverage—specific coverage will vary between employers.

EMPLOYMENT AND OUTLOOK

Materials engineers held 25,100 jobs in 2020. Employment of materials engineers is projected to grow 8 percent from 2020 to 2030, about as fast as the average for all occupations.

About half of all materials engineers work in manufacturing industries, where they develop and create new materials and manufacturing processes. Developing new materials or improving existing ones often requires collaboration with various engineering disciplines, such as mechanical, electrical, and chemical. As demand continues to increase for materials use, demand for employment also will increase. A greater focus on environmental sustainability also may create demand for materials engineers.

Related Occupations

- Aerospace Engineer
- Architectural/Engineering Manager
- Bioengineer/Biomedical Engineer
- Chemical Engineer
- Chemist/Materials Scientist
- Electrical/Electronics Engineer
- Mechanical Engineer
- Physicist/Astronomer
- Sales Engineer

MORE INFORMATION

Accreditation Board for Engineering and Technology, Inc. (ABET)
415 N. Charles Street
Baltimore, MD 21201
1.410.347.7700
www.abet.org

The American Ceramic Society (ACerS)
550 Polaris Parkway, Suite 510
Westerville, OH 43082
866.721.3322
ceramics.org

The American Institute of Mining, Metallurgical, and Petroleum Engineers (AIME)
12999 E. Adam Aircraft Circle
Englewood, CO 80112
303.948.4255
council@craftcouncil.org
www.aimehq.org

American Society for Engineering Education (ASEE)
1818 N Street, NW, Suite 600
Washington, DC 20036
202.331.3500
www.asee.org

ASM International
9639 Kinsman Road
Materials Park, OH 44073-0002
440.338.5151
memberservicecenter@asminternational.org
www.asminternational.org

Materials Research Society (MRS)
506 Keystone Drive
Warrendale, PA 15086-7537
724.779.3003
www.mrs.org

The Minerals, Metals & Materials Society (TMS)
5700 Corporate Drive, Suite 750
Pittsburgh, PA 15237
800.759.4867
tmsgeneral@tms.org
www.tms.org

National Action Council for Minorities in Engineering (NACME)
One North Broadway, Suite 601
White Plains, NY 10601
212.691.1051
info@nacme.org
www.nacme.org

The National Council of Examiners for Engineering and Surveying (NCEES)
200 Verdae Boulevard
Greenville, SC 29607
800.250.3196
ncees.org
www.arts-accredit.org

National Society of Professional Engineers (NSPE)
1420 King Street
Alexandria, VA 22314
888.285.6773
www.nspe.org

Society for Mining, Metallurgy & Exploration (SME)
12999 E. Adam Aircraft Circle
Englewood, CO 80112
720.574.1256
cs@smenet.org
www.smenet.org

Kristina Domizio & Micah L. Issitt

Conversation With...

KIM K. DE GROH

Senior Materials Research Engineer
NASA Glenn Research Center, Cleveland, OH
Materials Engineer, 27 years

What was your individual career path in terms of education, entry-level job, or other significant opportunity?

I became interested in the space program as a little girl when my father, a bio-medical engineer, shared his interest with me. During the Apollo 11 mission, my family watched together as Neil Armstrong took the first step onto the moon.

At Michigan State University (MSU), I decided to pursue a bachelor's in materials science because it encompassed various fields of interest to me: math, chemistry, atomic physics and art. My first technical job, a summer internship, taught me basic metallography skills and I realized I enjoyed research. I got my master's in materials science with an emphasis on electron microscopy—which I love. I trained in scanning, analytical and transmission electron microscopy, which helped me obtain my job at NASA.

As a grad student, I got an internship at NASA Glenn Research Center in Ohio. That summer, I met my future husband, Henry de Groh III, a NASA materials research engineer who was flying fascinating microgravity experiments on the shuttle. This confirmed my dream of someday flying materials experiments in space.

After earning my master's in 1987, I got a job as a contractor at NASA Glenn. One year later, NASA hired me as a materials research engineer.

I've had numerous materials experiments flown in space on the shuttle and the Russian space station Mir. As a senior materials research engineer in the Space Environment and Experiments Branch, I assess how space affects external spacecraft materials. I'm the principal investigator for 13 experiments flown on the exterior of the International Space Station (ISS) as part of the Materials International Space Station Experiment. I had a direct impact on the insulation chosen for replacement solar arrays placed on the Hubble Space Telescope during the first servicing mission and have analyzed insulation retrieved from Hubble during each of the five servicing missions.

What are the most important skills and/or qualities for someone in your profession?

Analytical, math and problem-solving skills, along with the ability to speak and write well are all important.

What do you wish you had known going into this profession?

I could have benefited from courses in statistics, error analyses and electrical engineering.

Are there many job opportunities in your profession? In what specific areas?

Many materials engineers work in metal production and processing, but also in the manufacture of electronic components, transportation equipment, and industrial equipment, as well as in research and testing services. Others work for federal and state agencies, such as NASA. The Bureau of Labor Statistics says materials engineers are in demand in biomedicine, working, for instance, on new materials for medical implants.

How do you see your profession changing in the next five years? What role will technology play in those changes, and what skills will be required?

Technology has led to drastic changes in the way I conduct and report research. Early in my career, we took Polaroid photos of microscope images and made our own viewgraph transparencies for presentations. Interaction with other researchers usually was limited to face-to-face meetings. Now I can operate a scanning electron microscope in one building and access the images from my office. I carry my presentations around on a flash drive. And e-mail has become a critical communication tool. I collaborate with researchers at NASA centers across the country and with space experts from around the world, such as at the European Space Agency and Japan Aerospace Exploration Agency. In the future, we'll likely see new materials discoveries, particularly in nanotechnology, composites, polymers and biomedicine.

What do you enjoy most about your job? What do you enjoy least about your job?

It's exciting to have my experiments taken into space, and even placed outside the International Space Station by an astronaut during a spacewalk. After a certain amount of space exposure, the experiments are brought back to Earth and I get to test how well the samples survived in the space environment. I write about the results and my data is used by spacecraft designers to build more durable spacecraft!

I have enjoyed presenting my results at conferences and traveling throughout the United States and to Canada, Europe and Japan to attend space conferences. But I think the best part of my job is the people I have met: bright and energetic students, fellow researchers, and astronauts who have helped with my projects.

What I enjoy least are the non-technical tasks: paperwork, conference participation approvals, etc.

Can you suggest a valuable "try this" for students considering a career in your profession?

Job shadowing provides great exposure to daily work. Internships can help you determine if you truly enjoy the work and whether you prefer lab work or something else, such as numerical modeling. Internships also allow you to showcase your abilities.

I recommend taking a course such as "Introduction to Careers in Engineering." My husband and I both took one. It helped me settle upon materials science.

This interview was originally published in 2015.

Mechanical Engineer

Snapshot

Career Cluster(s): Architecture & Construction; Science, Engineering, Technology & Mathematics

Interests: Science; Technology; Mechanics; Design

Earnings (Yearly Average): $90,160

Employment & Outlook: As Fast As Average

OVERVIEW

Sphere of Work

Mechanical engineers research, design, develop, build, and test mechanical and thermal sensors and devices, including tools, engines, and machines. Mechanical engineering is one of the broadest engineering fields. Mechanical engineers design and oversee the manufacture of many products ranging from medical devices to new batteries. Mechanical engineers design power-producing machines, such as electric generators, internal combustion engines, and steam and gas turbines, as well as power-using machines, such as refrigeration and air-conditioning systems. Mechanical engineers design other machines inside buildings, such as elevators and escalators. They also design material-handling systems, such as conveyor systems and automated transfer stations.

Photo via iStock/monkeybusinessimages. [Used under license.]

Work Environment

Mechanical engineers generally work in offices. They may occasionally visit worksites where a problem or piece of equipment needs their personal attention. In most settings, they work with other engineers, engineering technicians, and other professionals as part of a team.

Profile

Working Conditions: Work Indoors
Physical Strength: Light Work
Education Needs: Bachelor's Degree
Licensure/Certification: Not Required for Entry-level Positions
Opportunities for Experience: Internship; Apprenticeship; Part-time Work
Interest Score: IRC

Occupation Interest

Individuals interested in pursuing a career as a mechanical engineer should be skilled in scientific fields, as well being creative and innovative when it comes to discovering solutions when faced with problems. Mechanical engineers oftentimes spend their time troubleshooting designs they've created, so patience is required for times of trial and error.

A Day in the Life—Duties and Responsibilities

Mechanical engineers perform diverse tasks, including designing, programming, and project managing, depending on their credentials and the employers who need their services. These engineers apply their ingenuity in such areas as structural analysis, mechanical design, and systems manufacturing for corporations such as Boeing and government agencies such as the National Aeronautics and Space Administration (NASA).

Specific project goals determine how mechanical engineers conduct their daily tasks, whether in office buildings, factories, or other sites. Many devote part of their time to meeting with managers, colleagues, and clients to establish how a project should proceed. Mechanical engineers are often responsible for determining the budgets for specific projects. Sometimes these engineers are tasked with evaluating how their innovations might affect the environment and implementing measures to limit emissions and energy consumption. They consult with colleagues, both mechanical engineers and representatives of other specialties, when necessary. Mechanical engineers may also create computer software to aid them in assessing their manufacturing systems. Many mechanical engineers work to improve the safety and quality standards for new technology set by the American Society of Mechanical Engineers (ASME).

Duties and Responsibilities

- Analyzing problems to see how mechanical and thermal devices might help solve a particular problem
- Designing or redesigning mechanical and thermal devices or subsystems, using analysis and computer-aided design (CAD) software
- Investigating equipment failures or difficulties to diagnose faulty operation and recommending remedies
- Developing and testing prototypes of devices they design
- Analyzing test results and changing the design or system as needed
- Overseeing the manufacturing process for the device

Mechanical engineers working for universities are responsible for preparing engineering-specific curricula and guiding students in their course choices and research projects. Engineers who patent devices or processes may spend time seeking investors to fund their research and help them manufacture and sell their products. Some mechanical engineers assist government officials, including members of the U.S. Congress and the White House Office of Science and Technology, by advising leaders regarding mechanical engineering issues that affect federal policies addressing safety, environmental, or economic concerns.

OCCUPATION SPECIALTIES

Auto Research Engineer

Auto research engineers seek to improve the performance of cars. These engineers work to improve traditional features of cars such as suspension, and they also work on aerodynamics and new possible fuels.

Cartoonist

Cartoonists create simplified or exaggerated drawings to visually convey political, advertising, comic, or sports concepts. Some cartoonists work with others who create the idea or story and write captions. Others create plots and write captions themselves. Most cartoonists have humorous, critical, or dramatic talent, in addition to drawing skills.

Heating and Cooling Systems Engineer

Heating and cooling systems engineers work to create and maintain environmental systems wherever temperatures and humidity must be kept within certain limits. They develop such systems for airplanes, trains, cars, schools, and even computer rooms.

Robotic Engineer

Robotic engineers plan, build, and maintain robots. These engineers plan how robots will use sensors for detecting things based on light or smell, and they design how these sensors will fit into the designs of the robots.

WORK ENVIRONMENT

Immediate Physical Environment

Mechanical engineers generally work in offices. They may occasionally visit worksites where a problem or piece of equipment needs their personal attention. In most settings, they work with other engineers, engineering technicians, and other professionals as part of a team.

Human Environment

Mechanical engineers typically interact with fellow coworkers on a daily basis, whether it is working under the supervision of a senior engineer, or being the more experienced engineer overlooking the work of lower-level engineers. Additionally, mechanical engineers frequently work collaboratively in team projects with their colleagues, as well as engineering technicians and other professionals.

Technological Environment

Like other engineers, mechanical engineers use computers extensively. Mechanical engineers are routinely responsible for the integration of sensors, controllers, and machinery. Computer technology helps mechanical engineers create and analyze designs, run simulations and test how a machine is likely to work, interact with connected systems, and generate specifications for parts.

EDUCATION, TRAINING, AND ADVANCEMENT

High School/Secondary

For those in high school interested in becoming a mechanical engineer, enrolling in any available science and mathematics course to sharpen their skills will benefit them. Furthermore, if a school has any technology based or engineering classes, then it's advised to participate in those as well. Participation in science and engineering based extracurricular clubs will offer additional experience, as well as taking part in any science fair or robotics competition.

Suggested High School Subjects

- Algebra
- Applied Math
- Biology
- Calculus
- Chemistry

- Computer Science
- English
- Environmental Science
- Geometry
- Physics
- Pre-Calculus
- Trigonometry

Related Career Pathways/Majors

Architecture & Construction Career Cluster
- Design/Pre-Construction Pathway

Science, Technology, Engineering & Mathematics Career Cluster
- Engineering & Technology Pathway

Transferable Skills and Abilities

Creativity
- Designing and building complex pieces of equipment and machinery; a creative mind is essential

Listening Skills
- Listening to fellow architects and computer scientists when working on a project together
- Analyzing different approaches made by other experts to complete the task at hand

Math Skills
- Using the principles of calculus, statistics, and other advanced subjects in math for analysis, design, and troubleshooting in their work

Mechanical Skills
- Applying basic engineering concepts and mechanical processes to the design of new devices and systems

Problem-solving Skills
- Having good problem-solving skills to take scientific principles and discoveries and using them to design and build useful products

Postsecondary

Mechanical engineers typically need a bachelor's degree in mechanical engineering or mechanical engineering technologies. Mechanical engineering programs usually include courses in mathematics and life and physical sciences, as well as engineering and design. Mechanical engineering technology programs focus less on theory and more on the practical application of engineering principles. They may emphasize internships and co-ops to prepare students for work in industry.

Some colleges and universities offer five-year programs that allow students to obtain both a bachelor's and a master's degree. Some five-year or even six-year cooperative plans combine classroom study with practical work, enabling students to gain valuable experience and earn money to finance part of their education.

Related College Majors
- Aerospace Engineering
- Applied Physics
- Civil Engineering
- Industrial Engineering
- Materials Engineering
- Mechanical Engineering
- Metallurgical Engineering

- Mining & Mineral Engineering
- Naval Architecture & Marine Engineering
- Nuclear Engineering
- Robotics Engineering

Adult Job Seekers

People interested in pursuing a career in mechanical engineering should seek out a university or community college that offers courses or a degree in mechanical engineering; otherwise, it is unlikely to attain an entry-level position as bachelor's degrees are necessary. Additionally, internships or apprenticeships also help and individual gain experience with the various technologies and work environment of a professional mechanical engineer.

Professional Certification and Licensure

Licensure is not required for entry-level positions as a mechanical engineer. A Professional Engineering (PE) license, which allows for higher levels of leadership and independence, can be acquired later in one's career. Licensed engineers are called professional engineers (PEs). A PE can oversee the work of other engineers, sign off on projects, and provide services directly to the public. State licensure varies from state to state so be sure to check the requirements of where one may work or live.

Additional Requirements

Several states require engineers to take continuing education to renew their licenses every year. Most states recognize licensure from other states, as long as the other state's licensing requirements meet or exceed their own licensing requirements. Several professional organizations offer a variety of certification programs for engineers to demonstrate competency in specific fields of mechanical engineering.

EARNINGS AND ADVANCEMENT

Median annual earnings of mechanical engineers were $90,160 in 2020. The lowest 10 percent earned less than $58,410, and the highest 10 percent earned more than $141,060.

Generally, the pay of mechanical engineering has many variables that can impact the earnings of an individual. From the location of an employer to the size and type of business, an individual's education level, skill set, and job responsibilities

Fast Fact

Engineers are involved in most aspects of theme park design, from the rides to aspects such as crowd control. Source: ie-services.com

Photo via Photo via iStock/Imgorthand. [Used under license.]

can often reflect one's earnings. Most mechanical engineers work full time and some work more than 40 hours a week.

A PhD is essential for engineering faculty positions in higher education, as well as for some research and development (R&D) programs. Mechanical engineers may earn graduate degrees in engineering or business administration to learn new technology, broaden their education, and enhance their project management skills. Mechanical engineers may become administrators or managers after gaining work experience.

EMPLOYMENT AND OUTLOOK

Mechanical engineers held 299,200 jobs in 2020. Employment of mechanical engineers is projected to grow 7 percent from 2020 to 2030, about as fast as the average for all occupations.

Mechanical engineers work in many industries and on many types of projects. As a result, employment growth for these workers will vary by industry. As manufacturing processes incorporate more complex automation machinery, mechanical engineers are expected to be needed to help plan for and design this equipment. In automotive manufacturing, these engineers will play a key role in improving the range and performance of hybrid and electric cars. However, employment declines in some industries may temper overall employment growth of mechanical engineers.

Related Occupations

- Drafter
- Materials Engineer
- Mathematician/Statistician
- Mechanical Engineering Technologist/Technician
- Natural Science Manager
- Nuclear Engineer

- Petroleum Engineer
- Physicist/Astronomer
- Sales Engineer

MORE INFORMATION

American Society for Engineering Education (ASEE)
1818 N Street, NW, Suite 600
Washington, DC 20036
202.331.3500
www.asee.org

The American Society of Heating, Refrigerating and Air-Conditioning Engineers (ASHRAE)
180 Technology Parkway, NW
Peachtree Corners, GA 30092
404.636.8400
www.ashrae.org

The American Society of Mechanical Engineers (ASME)
Two Park Avenue
New York, NY 10016-5990
800.843.2763
CustomerCare@asme.org
www.asme.org

Engineers Without Borders USA
1031 33rd Street, Suite 210
Denver, CO 80205
303.772.2723
info@ewb-usa.org
www.ewb-usa.org

The National Council of Examiners for Engineering (NCEES)
200 Verdae Boulevard
Greenville, SC 29607
800.250.3196
ncees.org

The National Society of Black Engineers (NSBE)
205 Daingerfield Road
Alexandria, VA 22314
703.549.2207
adaa@artdealers.org
www.nsbe.org

The National Society of Professional Engineers (NSPE)
1420 King Street
Alexandria, VA 22314-2794
888.285.6773
www.nspe.org

Society of American Military Engineers (SAME)
1420 King Street, Suite 100
Alexandria, VA 22314
703.549.3800
www.same.org

Society of Hispanic Professional Engineers (SHPE)
13181 Crossroads Parkway N, Suite 220
City of Industry, CA 91746
www.shpe.org

Society of Women Engineers (SWE)
130 E. Randolph Street, Suite 4500
Chicago, IL 60601
1.312.596.5223
hq@swe.org
swe.org

Kristina Domizio & Elizabeth D. Schafer

Conversation With...
DAVID J. REESE

Practice Leader, Mechanical
KCI Technologies, Sparks, MD
Mechanical Engineer, 35 years

What was your individual career path in terms of education/training, entry-level job, or other significant opportunity?

I attended a technical high school that had an engineering curriculum that I enjoyed and excelled in, and I was good at math and enjoyed solving problems. Originally, I had intended to be an architect, but by the time I graduated from high school, I decided that engineering was a better fit for me.

In college, I did a work-study co-op program through Virginia Tech, where you get the practical experience of working for a company. I don't know that I learned a whole lot, technically, in the workplace, but I got a working knowledge of what being a consulting engineer entails. Going to college via that program meant it took five years to earn an undergraduate degree, but I also earned money to pay for school. After graduation, I went to work at the same company; they knew me and I knew them.

I was drawn to mechanical engineering because it was a little more challenging than some of the other disciplines. I took the exam that's the first step toward sitting for your Professional Engineering Exam and getting licensed when I was a college senior.

Early in my career, I became very good at designing chilled water piping and pumping systems. The industry has become quite dynamic in recent years for requiring sustainability and energy requirements. It's made us have to work harder and smarter, and it's definitely an improvement, from a technical standpoint, from what we were doing when I first started out.

I came in to my current employer as a senior mechanical engineer and have since been promoted to lead the mechanical engineering practice. This is the eighth company I've worked for during my career, and that's pretty typical for the industry with the exception of those who may stay in one company for a whole career.

What are the most important skills and/or qualities for someone in your profession?

Number one is good communication. You're dealing with people all the time.

What do you wish you had known going into this profession?

In school, they don't really teach you a practical knowledge of building codes. When I started out, every jurisdiction had its own code. They've been going toward a standardized code but all jurisdictions are still allowed to have their own requirements on top of standard codes. A better working knowledge of building codes would have been helpful.

Are there many job opportunities in your profession? In what specific areas?

There are many jobs available in the mechanical engineering profession such as consulting, as I'm doing, and also in sales, or even contracting and construction.

How do you see your profession changing in the next five years, what role will technology play in those changes, and what skills will be required?

Energy-efficient systems and their design has been a huge factor recently and it's going to have a major impact on the industry. I think more than anything, you will see this in the equipment manufacturing realm of this industry because we need to find equipment that's more efficient, and efficiency is being ramped up. A couple of recent examples include one, known as a variable refrigerant flow system (VRF) that's big in Asia; these systems are highly efficient. Another that is relatively new to industry is the Energy Wheel, where you pre-treat outdoor air using exhaust air from a building.

What do you enjoy most about your job? What do you enjoy least about your job?

What I most enjoy is the people in the industry. I enjoy working with people and teaching young people what I've learned throughout my career. I least enjoy bureaucracy and paperwork.

Can you suggest a valuable "try this" for students considering a career in your profession?

I'm Chairman of the Baltimore chapter of ASHRAE, the American Society of Heating, Refrigerating and Air-Conditioning Engineers (www.ashrae.org), and our chapter has a student challenge every year for local high school students. It's a teamwork type of exercise where participants actually get to do some design work for an air conditioning system for a commercial application of one sort or another. See if you can find a challenge such as that; participating in it would give you a real good clue if this is for you.

This interview was originally published in 2015.

Conversation With...
SARAH MCELRATH

Senior Mechanical Associate
Nuclear Power Technologies

What was your career path?
My career path started while I was in college and I interned at a nuclear power plant. After I graduated with a degree in mechanical engineering, I went to work full time at the plant as a thermal performance engineer in the Systems Plant Engineering Group. During this time I attended graduate school parttime and received my master's degree in mechanical engineering with a focus in thermal sciences. I had been working at the plant about four years when I got a job as a design engineer for an engineering firm. The group I am in specializes in nuclear power and I currently do design changes for two nuclear power plants in New Jersey. My former experience as a thermal performance engineer helps in this role because it gave me a broad view of the overall plant operation while touching on individual plant equipment and performance. An example of this is the feedwater heaters. As a thermal performance engineer I was responsible for monitoring performance, and now as a design engineer I am currently the responsible engineer for a project to replace the feedwater heaters at a plant now. My former experience helps me to have a practical perspective for the job I am trying to do now and gives me insight as to how the end user might want the heater to perform.

What are three pieces of advice you would offer someone interested in your profession?
1. The first piece of advice is to have an open mind. When I was in college I had no idea what career path I would like to take. I met a recruiter for the nuclear plant at a job fair and interviewed for the intern position. I grew up near a nuclear plant and never considered that as a career option. I found out through my internship that the industry was very interesting and I enjoyed the power generation technology and how it offers an opportunity to continue to learn things every day.

2. The second piece of advice is to find something that you enjoy. It is nice to get a paycheck but it means so much more to be contributing to an industry and profession that I care about.

3. The third piece of advice is to have a learning attitude every day. I continue to challenge myself with new projects and try to diversify my experiences. Even though I am done with school I am open to learning and the more diversified experience you have the more opportunities that may open up in the future.

What paths for career advancement are available to you?

In both working at a utility and at an engineering firm there are two main paths available for people with an engineering background: technical and management. You can continue to gain technical experience and advance through this career path. You could become a subject matter expert on a piece of equipment for example or a consultant that may provide advice to power plants trying to solve a difficult issue. The second career path is management. Following this career path puts you in a position of managing groups of people or even advancing to a position like plant manager, responsible for the entire plant. I have been working about six years and people around my age typically decide what track they would like to follow.

This interview was originally published in 2013.

Mechanical Engineering Technologist/Technician

Snapshot

Career Cluster(s): Architecture & Construction; Science, Technology, Engineering & Mathematics

Interests: Science; Mechanics; Design; Technology

Earnings (Yearly Average): $58,230

Employment & Outlook: As Fast as Average

OVERVIEW

Sphere of Work

Mechanical engineering technicians and technologists help mechanical engineers design, develop, test, and manufacture tools, engines, machines, and other devices. They may make sketches and rough layouts, record and analyze data, and report their findings. Mechanical engineering technicians also help mechanical engineers manufacture industrial machinery and other equipment.

Work Environment

Majority of mechanical engineering technicians work in factories or research and development (R&D) laboratories. Throughout the day, they will be participating in many tests of project designs and as a result, there is a potential risk to being exposed to hazards. This ranges from equipment, chemicals, or toxic materials. Proper safety proce-

Photo via iStock/Morsa Images. [Used under license.]

dures must be followed when handling machinery and equipment to keep the risk of injuries low.

Profile

Working Conditions: Mainly Indoors
Physical Strength: Light Work
Education Needs: On-the-Job Training; Associate's Degree; Bachelor's Degree
Licensure/Certification: Not Required
Opportunities for Experience: Internship; Apprenticeship
Interest Score: RI

Occupation Interest

Individuals interested in pursuing a career as a mechanical engineering technician should have a strong interest in both science and mathematics, as an individual in the field will be working on projects on a daily basis that require good math skills for analysis, design and troubleshooting of any issues that may arise. It is also important than an individual has great problem-solving skills as they will be working alongside a senior engineer on their designs and eventually will be required to overcome any problems that emerge throughout the development and testing period of a project.

A Day in the Life—Duties and Responsibilities

Mechanical engineering technicians typically install, troubleshoot, and maintain designs created by engineers. They may work under the direction of engineers or engineering technologists. Mechanical engineering technicians typically help engineers improve designs or incorporate new technology. They may be team leaders, instructing lower-level technicians on installing equipment, systems, or structures. Some mechanical engineering technicians test and inspect machines and equipment or work with engineers to eliminate production problems. For example, they may help test products by setting up instruments for vehicle crash tests.

Duties and Responsibilities

- Evaluating specifications in design drawings prior to adding or replacing tools
- Preparing layouts and drawings of the assembly process and parts to be made, usually using three-dimensional (3D) design software
- Recommending cost-effective changes in equipment design to improve reliability and safety
- Reviewing instructions and blueprints to ensure that project plans follow test specifications and procedures and meet objectives
- Planning, producing, and assembling mechanical parts for products, such as industrial equipment
- Setting up and conducting tests of complete units and their components, and recording results
- Comparing test results with design specifications and with test objectives and recommending changes, if needed, in products or test methods
- Making calculations for business estimates, such as of labor costs and equipment lifespan

Throughout the day, mechanical engineering technicians are frequently working alongside and collaboratively with mechanical engineers by helping design, develop, test, and manufacture a project. This could range from many devices, including engines, machines, or tools. Another aspect of a mechanical engineering technician's day is spent creating sketches and rough layouts of designs, perhaps building off an idea of a senior mechanical engineer. Additionally, it is their job the record and analyze the data found during test trials of projects, as well as reporting it to their supervisors and fellow technicians who may be working on the same project.

WORK ENVIRONMENT

Immediate Physical Environment

Mechanical engineering technicians work primarily in factories or in R&D labs. Some technicians, particularly those working in factories, are exposed to hazards from equipment, chemicals, or toxic materials. However, their risk of injury or illness is low if they follow procedures. Because mechanical engineering technicians work with machines of all types, they must take safety precautions in their workspace.

Human Environment

A mechanical engineering technician often works under the direct supervision of a senior engineer, whose designs and plans are frequently what they are working to adapt into a functioning product. Also, technicians would be working alongside other mechanical engineering technicians, some of whom may also be working on the same project. In this instance, a technician will require extensive interpersonal communication skills as they would be working with others quite often throughout the day.

Technological Environment

Mechanical engineering technicians may need to operate machinery such as drill presses, grinder, and engine lathes when it comes to properly testing equipment from its design phase to manufacturing. Alongside the testing process of machines and designs, technicians may use computer-aided design software to design and develop any projects given to them by senior engineers. Due to mechanical engineering technicians frequently being around industrial machinery and equipment, proper safety precautions must be met.

EDUCATION, TRAINING, AND ADVANCEMENT

High School/Secondary

High school students interested in becoming a mechanical engineering technician should take classes in math, science, and computer science. Additionally, participation in extracurricular clubs involving computer science, science, or robotics would also be beneficial to gain experience.

Suggested High School Subjects

- Algebra
- Applied Math
- Biology
- Calculus
- Computer Science
- English
- Geometry
- Physics
- Trigonometry

Related Career Pathways/Majors

Architecture & Construction Career Cluster

- Design/Pre-Construction Pathway

Science, Technology, Engineering & Mathematics Career Cluster

- Engineering & Technology Pathway

Transferable Skills and Abilities

Communication Skills

- Exhibiting an ability to clearly explain, both orally and in writing, the need for changes in designs or test procedures

Detail-oriented

- Making precise measurements and keeping accurate records for mechanical engineers

Math Skills

- Using mathematics for analysis, design, and troubleshooting

Mechanical Skills

- Applying theory and instructions from engineers by making new components for industrial machinery of equipment
- Operating machinery such as drill presses, grinders, and engine lathes

Problem-solving Skills

- Helping mechanical engineers bring their plans and designs to life
- Helping engineers overcome problems that emerge throughout the development or manufacturing

Postsecondary

Mechanical engineering technicians typically need an associate's degree or a certificate from a community college or vocational-technical school in order to enter the occupation. Employers may prefer to hire candidates who have completed a bachelor's degree or whose programs are accredited by an organization such as ABET (Accreditation Board for Engineering and Technology, Inc.).

Related College Majors

- Chemistry
- Civil Engineering
- Materials Science & Engineering
- Mechanical Engineering

Adult Job Seekers

Individuals interested in a career as a mechanical engineering technician should be proficient in both scientific and mathematical fields. An ability to troubleshoot problems in a design and work through difficult problems that may arise during the testing and manufacturing phase is required, as is patience for many trial-and- error runs. Furthermore, a person must be innovative in order to further improve on designs given by a senior mechanical engineer and to make them work correctly.

Professional Certification and Licensure

The National Institute for Certification in Engineering Technologies (NICET) offers optional credentials through its technician certification programs. For example, mechanical engineering technologists and technicians who examine water-based fire protection systems may obtain certification in fire sprinkler layout or fire sprinkler inspection and testing. An engineering technologist's or technician's supervisor usually must verify the candidate's performance competency as part of the certification process.

Additional Requirements

Individuals should have an inclination toward science and mathematics since both are skills that a mechanical engineering technician will use on a daily basis. Additionally, working under stress from complex problems that require an immediate response to fix is another requirement that many should be prepared to handle.

EARNINGS AND ADVANCEMENT

Median annual earnings of mechanical engineering technologists and technicians were $58,230 in 2020. The lowest 10 percent earned less than $36,200, and the highest 10 percent earned more than $91,320. Technicians who obtain appropriate education or certification may advance to become technologists.

EMPLOYMENT AND OUTLOOK

Mechanical engineering technologists and technicians held 40,400 jobs in 2020. Employment is expected to show slow change through the year 2030. Because mechanical engineering technologists and technicians are directly supervised by mechanical engineers, their employment growth will be driven by many of the same factors driving employment growth for mechanical engineers. These factors include demand for engineering services in the motor vehicle parts and machinery manufacturing industries.

Companies may contract out the work of mechanical engineering technologists and technicians to consulting firms, which should lead to employment growth in this industry. Contracting for this work allows companies to obtain engineering services at a lower cost than employing in-house technologists and technicians.

Related Occupations

- Drafter
- Industrial Engineering Technologist/Technician
- Machinist/Tool/Die Maker
- Mechanical Engineer

MORE INFORMATION

Accreditation Board for Engineering and Technology, Inc. (ABET)
415 N. Charles Street
Baltimore, MD 21201
410.347.7700
www.abet.org

American Society for Engineering Education (ASEE)
1818 N Street, NW, Suite 600
Washington, DC 20036
202.331.3500
www.asee.org

The American Society of Mechanical Engineers (ASME)
Two Park Avenue
New York, NY 10016-5990
800.843.2763
www.asme.org

Technology Student Association (TSA)
1904 Association Drive
Reston, VA 20191-1540
703.860.9000
general@tsaweb.org
tsaweb.org

Kristina Domizio

Mining/Geological Engineer

Snapshot

Career Cluster(s): Science, Technology, Engineering & Mathematics

Interests: Earth Science; Geology; Mining; Technology; Environmental Science

Earnings (Yearly Average): $93,800

Employment & Outlook: Slower Than Average Growth Expected

OVERVIEW

Sphere of Work

Mining and geological engineers design mines to safely and efficiently remove minerals such as coal and metals for use in manufacturing and utilities.

Work Environment

Some mining and geological engineers work for engineering firms in offices, while others work on-site in remote locations to monitor mining and geological operations.

Photo via iStock/olli0815. [Used under license.]

Occupation Interest

Mining and geological engineers have a dual interest in engineering and geology, and seek to apply their ingenuity in a setting that benefits mining operations and geological excavations. They are responsible for the safe and environmentally friendly extraction of materials from the earth, and continually work to improve techniques in the field.

Profile

Working Conditions: Both Inside and Outside
Physical Strength: Medium to Heavy Work
Education Needs: Bachelor's Degree; Master's Degree
Licensure/Certification: Varies
Opportunities for Experience: Laboratory Work; Fieldwork
Interest Score: IRE

A Day in the Life—Duties and Responsibilities

Duties differ by occupation. Geological engineers search for mineral deposits and evaluate possible sites. Once a site is identified, they plan how the metals or minerals will be extracted in efficient and environmentally sound ways.

Mining engineers often specialize in one particular mineral or metal, such as coal or gold. They typically design and develop mines and determine the best way to extract metals or minerals to get the most out of deposits.

Some mining engineers work with geoscientists and metallurgical engineers to find and evaluate ore deposits. Other mining engineers develop new equipment or direct mineral-processing operations to separate minerals from dirt, rock, and other materials.

Duties and Responsibilities

- Designing open-pit and underground mines
- Supervising the construction of mine shafts and tunnels
- Devising methods for transporting minerals to processing plants
- Preparing technical reports for miners, engineers, and managers
- Monitoring mine production to assess the effectiveness of operations
- Providing solutions to problems related to land reclamation, water and air pollution, and sustainability
- Ensuring that mines are operated in safe and environmentally sound ways

Mining safety engineers use best practices and their knowledge of mine design to ensure workers' safety and to maintain compliance with state and federal safety regulations. They inspect the walls and roofs of mines, monitor the air quality, and examine mining equipment for possible hazards.

Engineers who hold a master's or a doctoral degree may teach engineering at colleges and universities.

WORK ENVIRONMENT

Immediate Physical Environment

Many work where mining operations are located, such as mineral mines or sand-and-gravel quarries, in remote areas or near cities and towns, to keep close watch on the progression of their designs. Others work in offices or on-site for oil and gas extraction firms or engineering services firms.

Most mining and geological engineers work full time and some work more than 40 hours a week. The remoteness of some mining locations gives rise to variable schedules and weeks during which they work more hours than usual.

Human Environment

Mining and geological engineers may interact with many different people throughout the course of their workdays, including other engineers, project managers, and site supervisors. If they work for an engineering firm, they will likely report to firm managers as well. They must be able to collaborate with others and work under potentially harsh conditions when visiting mining sites.

Technological Environment

Mining and geological engineers must be well-versed in the array of professional software required to perform their duties, including computer-aided design (CAD) programs, analytical and scientific software, enterprise resource planning (ERP) software, and database and presentation software. Knowledge of mining technology and procedures is also an asset, especially for those attending mining sites, as well as methods for environmental reclamation.

EDUCATION, TRAINING, AND ADVANCEMENT

High School/Secondary

High school students interested in studying mining or geological engineering should take classes in math and science.

Suggested High School Subjects

- Algebra
- Biology
- Calculus
- Chemistry

- Earth or Life or Physical Science
- Economics
- Entrepreneurship
- Geometry
- History
- Physics
- Pre-Calculus
- Statistics
- Trigonometry

Related Career Pathways/Majors

Science, Technology, Engineering and Mathematics Career Cluster

- Engineering & Technology Pathway
- Science & Mathematics Pathway

Transferable Skills and Abilities

Analytical Skills

- Taking many factors into account when evaluating new mine locations and designing facilities
- Planning for the restoration of the surrounding environment after operations end

Decision-making Skills

- Making decisions that influence many critical outcomes—from worker safety to mine production
- Anticipating problems and deal with them immediately

Logical-thinking Skills

- Putting work plans into a coherent, logical sequence

Math Skills

- Using the principles of calculus, trigonometry, and other advanced topics in math for analysis, design, and troubleshooting in their work

Problem-solving Skills

- Exploring for potential mines, planning their operations and mineral processing, and designing environmental reclamation projects
- Identifying and working toward goals, while solving problems along the way

Writing Skills

- Preparing reports and instructions for other workers
- Writing clearly so that others can easily understand their ideas and plans

Postsecondary

College students typically get a degree in an engineering field. Because relatively few schools offer programs in mining engineering or geological engineering, a degree in civil or environmental engineering or geoscience is often acceptable. Bachelor's degree programs in mining engineering typically include courses in geology, thermodynamics, and mine design and safety. Bachelor's degree programs in geological engineering typically include courses in geology, chemistry, and fluid mechanics. Both types of programs also include laboratory and fieldwork, along with academic study.

Employers may prefer to hire mining and geological engineering candidates who have graduated from a program accredited by a professional association such as the Accreditation Board for Engineering and Technology, Inc. (ABET).

Master's degree programs in mining and geological engineering typically are two-year programs and include coursework in specialized subjects, such as mineral resource development and mining regulations. Some programs require a written thesis for graduation.

Related College Majors

- Civil Engineering
- Environmental Engineering
- Geochemistry
- Geological Engineering
- Geology & Earth Sciences
- Geophysics & Seismology
- Geoscience Engineering
- Hydrology & Water Resources Science
- Ocean Engineering
- Oceanography
- Paleontology

Adult Job Seekers

Adults interested in becoming mining or geological engineers should have a background in related areas such as civil or environmental engineering, and be prepared to transition their careers. They might be able to gain practical experience by taking on mining or geological projects in their current jobs, if possible. Further coursework will likely also be required.

Professional Certification and Licensure

Licensure is not required for entry-level positions as a mining or geological engineer. A Professional Engineering (PE) license, which allows for higher levels of leadership and independence, can be acquired later in one's career. Licensed engineers are called professional engineers (PEs). A PE can oversee the work of other engineers, sign off on projects, and provide services directly to the public. State licensure generally requires:

- A degree from an ABET-accredited engineering program
- A passing score on the Fundamentals of Engineering (FE) exam
- Relevant work experience, typically at least four years
- A passing score on the Professional Engineering (PE) exam

The initial FE exam can be taken after one earns a bachelor's degree. Engineers who pass this exam are commonly called engineers in training (EITs) or engineer interns (EIs). After meeting work experience requirements, EITs and EIs can take the second exam, called the Principles and Practice of Engineering.

In several states, engineers must earn continuing education credits to keep their licenses. Most states recognize licenses from other states, provided that licensure requirements in the other states meet or exceed the first state's own requirements.

EARNINGS AND ADVANCEMENT

The median annual wage for mining or geological engineers in 2020 was $93,800. The lowest 10 percent earned less than $57,040, and the highest 10 percent earned more than $156,270.

Mining and geological engineers may receive benefits such as paid vacations, holidays, and sick days; life and health insurance; and retirement benefits, usually paid by the employer.

New mining and geological engineers usually work under the supervision of experienced engineers. In large companies, new engineers also may receive formal classroom or seminar-type training. As engineers gain knowledge and experience, they are assigned more difficult projects and they are given greater independence to develop designs, solve problems, and make decisions.

Engineers may advance to become technical specialists or supervise a staff or team of engineers and technicians. Some eventually become engineering managers or enter other managerial or sales jobs. In sales, an engineering background enables them to discuss a product's technical aspects and to assist in product planning, installation, and use.

EMPLOYMENT AND OUTLOOK

There were approximately 6,300 mining and geological engineers employed nationally in 2020. Employment of mining and geological engineers is projected to grow 4 percent from 2020 to 2030, slower than the average for all occupations.

Employment growth for mining and geological engineers will depend upon demand for coal, metals, and minerals. These resources are used in many products, from construction materials and cars to cell phones and computers. As companies look for ways to cut costs, they are expected to contract more services with engineering services firms, rather than employ engineers directly.

Famous First

The first metals to be unearthed were gold and copper, with some copper finds dating back to 8,700 BC.
Source: www.generalkinematics.com

Photo via iStock/assistantua. [Used under license.]

Related Occupations

- Architectural/Engineering Manager
- Civil Engineer
- Environmental Scientist/Specialist
- Geological/Hydrologic Technician
- Geoscientist
- Health & Safety Engineer
- Hydrologist
- Mechanical Engineer
- Petroleum Engineer
- Sales Engineer

MORE INFORMATION

Accreditation Board for Engineering and Technology, Inc. (ABET)
415 N. Charles Street
Baltimore, MD 21201
410.347.7700
www.abet.org

American Society for Engineering Education (ASEE)
1818 N Street, NW, Suite 600
Washington, DC 20036
202.331.3500
www.asee.org

National Council of Examiners for Engineering and Surveying (NCEES)
200 Verdae Boulevard
Greenville, SC 29607
800.250.3196
www.ncees.org

National Society of Professional Engineers (NSPE)
1420 King Street
Alexandria, VA 22314
888.285.6773
www.nspe.org

Society for Mining, Metallurgy, & Exploration (SME)
12999 E. Adam Aircraft Circle
Englewood, CO 80112
720.738.4085
cs@smenet.org
www.smenet.org

Technology Student Association (TSA)
1904 Association Drive
Reston, VA 20191-1540
703.860.9000
general@tsaweb.org
tsaweb.org

Stuart Paterson

Conversation With...
SARAH MORTON RUPERT

Research Geophysicist
Kansas Geological Survey, Lawrence, KS
Geotechnical Engineering, 8 years

What was your individual career path in terms of education/training, entry-level job, or other significant opportunity?
I was absolutely enthralled by my sixth grade earth science class. The next year, I did a research project on rocks and minerals and was completely sold. After taking high school physics, I knew I wanted to be a geophysicist.

I have a bachelor's degree in geoscience and a master's in civil engineering, with a specialization in geotechnical engineering. During my senior year as an undergrad at the University of Connecticut, I told my geophysics professor I was interested in near-surface seismic and seismic hazard research and he offered to take me on as a master's student. He arranged an interview with the chief of the U.S. Geological Survey (USGS) Office of Groundwater. I started as a "Volunteer for Science," researching surface wave seismic methods. I learned about theory and field methodology, reviewed case studies, and designed my own surveys. Later, I was offered an internship as "Student Trainee in Hydrology" with the USGS and a one-year contract as an environmental protection resource assistant with the Connecticut Geological Survey.

The chief of the USGC became my master's research advisor. He helped me find a research opportunity that turned into my thesis project, helping New England state geologists update their seismic hazard maps. I was able to support myself because the CT Geological Survey and USGS financially supported my research.

In 2013, I spent three separate weeks working with the Kansas Geological Survey (KGS), learning the principles of the downhole seismic method so that I could bring that information back to the USGS. The KGS Exploration Services section chief offered me a research assistantship as a PhD student at the University of Kansas. The KGS is world-renowned for their near-surface seismic program. To my delight, they hired me as a research geophysicist for six months to fill my time between defending my master's thesis and starting my PhD coursework. Today, I'm still a graduate research assistant at the KGS. I'm about to enter the fifth and (hopefully) final year of my PhD. I've been fortunate to tackle environmental and engineering problems all across the nation. This summer, I took a leave from the Kansas Geological Survey to do an internship as student trainee in civil engineering at the Army Corps of Engineers Engineering Development and Research Center (ERDC) in Vicksburg, MS. My plan is to continue working for a research institution, preferably with the federal government.

What are the most important skills and/or qualities for someone in your profession?
The most important skill is the ability to communicate with both a technical community and the general public, to make your science available to everyone. Communicating your science by writing technical papers isn't taught in universities; you learn through experience. Learn to write well and learn to take criticism; let comments be constructive, not destructive to your work. Learn to be creative: it will improve your presentations and help you think about your research differently. Learn to edit technical writing by becoming a reviewer for conference papers or journals.

What do you wish you had known going into this profession?
I wish I had known how important good, thorough note taking is. It's important to record enough information so that a survey or experiment can be repeated to obtain similar results. I thought I had a decent memory and felt that if I didn't have time to record something, I'd be able to remember it later. That has not always been the case. Great notes will reduce headaches when you're trying to remember how you got to a certain result.

Are there many job opportunities in your profession? In what specific areas?
There absolutely are opportunities in near-surface geophysics, especially in the environmental and engineering sectors, working in academia, government, and industry. An example of a private sector job would be working for a civil engineering/construction company doing site characterization. For instance, if a company wants to put in a foundation, a site characterization would determine if there are any soils that would fail under too much load or settle quickly if the proper precautions were not taken. The world is going to continue building infrastructure above and below ground, and geophysics helps us understand what lies underneath and make better-informed decisions. Near-surface geophysics can be used to monitor resources and changes in the landscape.

How do you see your profession changing in the next five years? What role will technology play in those changes, and what skills will be required?
I hope to see more engineering and environmental companies integrate geophysical techniques into their site assessments to improve decision making. As we become more aware of environmental problems such as climate change, I think the number of people going into geophysical, environmental, and engineering fields will increase.

With technology constantly advancing, we're finding new and more efficient ways to collect data, giving us access to remote areas. I foresee field equipment continuing to get lighter and more compact and therefore easier to transport. As hardware gets smaller and more economical, groups like universities may have better access to equipment and be able to conduct more surveys.

What do you enjoy most about your job? What do you enjoy least about your job?
I think my favorite part is the diversity of problems I work on. Also, so much of my work can help solve a problem or lead to better ways of handling problems. For instance, I worked on a

project where a contaminant plume was not migrating as originally anticipated. We identified faults and fractures that were diverting the flow, allowing the contamination to be remediated effectively.

My least favorite thing is how long the review process takes from the time you submit a manuscript or journal paper until you learn whether it has been accepted or rejected.

Can you suggest a valuable "try this" for students considering a career in your profession?

If you love being outside, hiking, or being in the water and the elements, a career in the earth sciences may be right for you. Try finding a rocks and mineral show and talk to people there. These events are usually hosted by a geoscience-related organization.

This interview was originally published in 2018.

Conversation With...

ANDREW P. SCHISSLER, PE

Principal, Schissler Engineering LLC
Adjunct Professor, Colorado School of Mines
Mining Engineer, SME PE Exam Coordinator
In the field, 42 years

What was your individual career path in terms of education/training, entry-level job, or other significant opportunity?

I always preferred technological and scientific subjects. I decided to go to the best science or technology university near my Colorado home, the Colorado School of Mines. I didn't have to declare a major until mid-sophomore year. I chose mining engineering because it combined elements of other engineering disciplines and had one of the highest starting salaries at that time.

After graduation, I worked for a large coal mining company in Fairmont, WV for two years and was in a management training program. Then, due to my aging parents, I moved back west, to Utah. My job as an engineer in a mine included mine design and related systems designs: water, power, haulage, mineral processing, and materials handling, as well as capital budget implementation.

I received my foreman certification during my second year of work. Well-grounded advancement comes from operations experience, and there's no better way to get it than as a first-line supervisor or by serving in a technical role at a mine.

I went on to move around and climb the ladder: as a chief engineer, superintendent of operations planning, mine manager, and manager of technical services. I worked for the fourth largest diversified mining company in the world from 1980 to 2000. Then I started and finished a Ph.D. that required a year on campus to focus and write my thesis. After graduation, I went to Penn State as an assistant professor. After three years, I decided to go back into industry as manager of engineering for a group of potash mines; potash is a critical agricultural fertilizer mineral. At that point, I was getting into my 50s and my wife's and my family is in Denver, so I made one more move—home, where I took an executive position. The 2008 financial crisis put my employer out of business so I went on to spend four years as a principal mining engineer for a major engineering company. Two years ago, I took on an operations manager role for another company, and then started my own consulting firm after that company had some restructuring a year ago.

In addition to being Principal Engineer of my own company, I am an adjunct professor at the Colorado School of Mines as well as the PE exam coordinator for the Society of Mining, Metallurgy and Exploration Inc.

One of my specialties has been rock mechanics. When dealing with rock, you need to know how strong it is. It's a science and practice of engineering but will always be challenging because rock is variable and not homogeneous.

What are the most important skills and/or qualities for someone in your profession?
You need the ability to listen and to show initiative.

What do you wish you had known going into this profession?
This industry is subject to economic cycles. During expansions, opportunity is high. During contractions, opportunity is low.

Are there many job opportunities in your profession? In what specific areas?
Overall, mining engineers will find a diverse range of opportunities. They may work at a mine, for a mining company, in mine construction or civil engineering, or in mining equipment manufacturing. You could also work as a technical service engineer, or in government with agencies like the Mine Safety Health Administration, the National Institute of Occupational Safety and Health, intelligence agencies analyzing underground (as in below the earth) foreign installations, or even as an officer in one of the armed forces. Mining Engineering is terrific pre-law or pre-med course of study as well.

How do you see your profession changing in the next five years, what role will technology play in those changes, and what skills will be required?
There's going to be continued re-definition of how engineering achieves the objective of extracting minerals. All disciplines of engineering are going to require additional training on the biological side and will need to consider societal interaction such as climate change. For example, all industrial and commercial enterprises incur what is called a carbon footprint. How do you calculate that? This will become the norm.

What do you enjoy most about your job? What do you enjoy least about your job?
I enjoy my diversity of assignments: this year, I've been to Australia, Spain and Peru as well as a few U.S. states. Each of my projects is different. I enjoy being my own boss, although—a small dislike—now I have to be my own accountant.

Can you suggest a valuable "try this" for students considering a career in your profession?
A Boy Scout can get a merit badge in mining. The mining industry is very supportive of this. In addition, many high school science departments bring in industry people to sponsor teams in science or engineering to perform an engineering project, which sometimes includes mining. Participating would be a good way to find an industry mentor. If you're already at a university, you might get a summer internship at a mining company.

This interview was originally published in 2015.

Nuclear Engineer

Snapshot

Career Cluster(s): Science, Technology, Engineering & Mathematics

Interests: Energy; Nuclear Power

Earnings (Yearly Average): $116,140

Employment & Outlook: Decline Expected

OVERVIEW

Sphere of Work

Nuclear engineers research and develop the processes, instruments, and systems used to derive benefits from nuclear energy and radiation.

Work Environment

Nuclear engineers typically work in offices; however, their work setting varies with the industry in which they are employed. Most nuclear engineers work full time.

Occupation Interest

Nuclear engineers strive to be at the forefront of engineering science. By its very nature, nuclear energy is extremely powerful, so becoming an engineer who harnesses this power takes years of study and a keen understanding of the scientific and engineering processes involved. Interests can intersect with others fields, such as aerospace engineering and the creation of medical devices.

Photo via iStock/PeterTG. [Used under license.]

A Day in the Life—Duties and Responsibilities

Nuclear engineers research and develop the processes, instruments, and systems used to derive benefits from nuclear energy and radiation. Many of these engineers find industrial and medical uses for radioactive materials—for example, in equipment used in medical diagnosis and treatment. Many others specialize in the development of nuclear power sources for ships or spacecraft.

Profile

Working Conditions: Inside Work
Physical Strength: Light Work
Education Needs: Bachelor's Degree; Master's Degree; Doctoral Degree
Licensure/Certification: Optional
Opportunities for Experience: Coursework; On-site Training
Interest Score: IRC

In addition, nuclear engineers are at the forefront of developing uses of nuclear material for medical imaging devices, such as positron emission tomography (PET) scanners. They also may develop or design cyclotrons, which produce a high-energy beam that the healthcare industry uses to treat cancerous tumors.

WORK ENVIRONMENT

Duties and Responsibilities

- Designing or developing nuclear equipment, such as reactor cores, radiation shielding, and associated instrumentation
- Directing operating or maintenance activities of operational nuclear power plants to ensure that they meet safety standards
- Writing operational instructions to be used in nuclear plant operation or in handling and disposing of nuclear waste
- Monitoring nuclear facility operations to identify any design, construction, or operation practices that violate safety regulations and laws
- Performing experiments to test whether methods of using nuclear material, reclaiming nuclear fuel, or disposing of nuclear waste are acceptable
- Taking corrective actions or ordering plant shutdowns in emergencies
- Examining nuclear accidents and gathering data that can be used to design preventive measures

Immediate Physical Environment

Nuclear engineers typically work in offices. However, their work setting varies with the industry in which they are employed. For example, those employed in power generation and supply work in power plants. Many work for the federal government and for consulting firms.

The majority of nuclear engineers work full time and some work more than 40 hours per week. Their schedules may vary with the industries in which they work.

Human Environment

Nuclear engineers work with others, including mechanical engineers and electrical engineers, and must be able to incorporate systems designed by these engineers into their own designs.

Technological Environment

Common programs used by nuclear engineers include development environment software and object or component-oriented development software. Analytical or scientific software, databases, and other common office programs may also be employed. Further, nuclear engineers require an understanding of how their technology is used in the field, be it a power-plant component or a medical device.

EDUCATION, TRAINING, AND ADVANCEMENT

High School/Secondary

Students interested in studying nuclear engineering should take high school courses in mathematics, such as algebra, trigonometry, and calculus; and science, such as biology, chemistry, and physics.

Suggested High School Subjects

- Algebra
- Biology
- Calculus
- Chemistry
- Civics
- Economics
- English
- Geography
- Geometry
- History
- Physics
- Statistics
- Trigonometry

Related Career Pathways/Majors

Science, Technology, Engineering & Mathematics Career Cluster

- Engineering & Technology Career Pathway
- Science & Mathematics Pathway

Postsecondary

Entry-level nuclear engineering jobs in private industry require a bachelor's degree. Some entry-level nuclear engineering jobs may require at least a master's degree or even a PhD.

Bachelor's degree programs consist of classroom, laboratory, and field studies in subjects such as mathematics and engineering principles. Most colleges and universities offer cooperative-education programs in which students gain work experience while completing their education.

Some universities offer five-year programs leading to both a bachelor's and a master's degree. A graduate degree allows an engineer to work as an instructor at a university or engage in research and development (R&D). Some five-year or even six-year cooperative-education plans combine classroom study with work, permitting students to gain experience and to finance part of their education.

Master's and PhD programs consist of classroom, laboratory, and research efforts in areas of advanced mathematics and engineering principles. These programs require the successful completion of a research study, usually conducted in conjunction with a professor, on a government or private research grant.

Programs in nuclear engineering are accredited by the Accreditation Board for Engineering and Technology, Inc. (ABET).

Related College Majors

- Architectural Engineering
- Civil Engineering
- Computer Engineering
- Construction Engineering
- Drafting & Design Engineering
- Electromechanical Engineering
- Electronics Engineering
- Energy Systems Technologies
- Environmental Control
- Industrial Production
- Mechanical Engineering
- Mining & Petroleum Engineering
- Nanotechnology
- Physics
- Quality Control

Transferable Skills and Abilities

Analytical Skills

- Identifying design elements to help build facilities and equipment that produce material needed by various industries

Communication Skills

- Working with other engineers and technicians; communicating effectively, both in writing and in person
- Working to invent new materials, advanced manufacturing techniques, and new applications in chemical and biomedical engineering

Detail-oriented

- Supervising the operation of nuclear facilities
- Paying close attention to what is happening at all times and ensuring that operations comply with all regulations and laws pertaining to the safety of workers and the environment

Logical-thinking Skills

- Ordering information logically and clearly so that others can follow their written information and instructions

Math Skills

- Using the principles of calculus, trigonometry, and other advanced topics in math for analysis, design, and troubleshooting in their work

Problem-solving Skills

- Anticipating problems before they occur and safeguarding against them

Fast Fact

Because of its high density, all used nuclear fuel produced by the U.S. over the last 60 years could fit on a football field at a depth of less than 10 yards.
Source: www.energy.gov

Photo via iStock/RelaxFoto.de. [Used under license.]

Professional Certification and Licensure

Licensure is not required for entry-level positions as a nuclear engineer. A Professional Engineering (PE) license, which allows for higher levels of leadership and independence, can be acquired later in one's career. Licensed engineers are called professional engineers (PEs). A PE can oversee the work of other engineers, sign off on projects, and provide services directly to the public. State licensure generally requires:

- A degree from an ABET-accredited engineering program
- A passing score on the Fundamentals of Engineering (FE) exam
- Relevant work experience, typically at least four years
- A passing score on the Professional Engineering (PE) exam

The initial FE exam can be taken after one earns a bachelor's degree. Engineers who pass this exam are commonly called engineers in training (EITs) or engineer interns (EIs). After meeting work experience requirements, EITs and EIs can take the second exam, called the Principles and Practice of Engineering.

Each state issues its own licenses. Most states recognize licensure from other states, as long as the licensing state's requirements meet or exceed their own licensure requirements. Several states require continuing education for engineers to keep their licenses.

Nuclear engineers can obtain licensing as a Senior Reactor Operator, a designation that is granted after an intensive, two-year, site-specific program. The credential, granted by the Nuclear Regulatory Commission, asserts that the engineer can operate a nuclear power plant within federal government requirements.

Additional Requirements

A newly hired nuclear engineer at a nuclear power plant usually must complete training on-site, in such areas as safety procedures, practices, and regulations, before being allowed to work independently. Training lasts from six weeks to three months, depending on the employer. In addition, these engineers must undergo continuous training every year to keep their knowledge, skills, and abilities current with laws, regulations, and safety procedures.

EARNINGS AND ADVANCEMENT

Earnings of nuclear engineers depend on their education and experience, employer, and the scope of projects overseen. Median annual earnings of nuclear engineers were $116,140 in 2020. The lowest 10 percent earned less than $74,800, and the highest 10 percent earned more than $185,550.

Nuclear engineers may receive benefits such as paid vacations, holidays, and sick days; life and health insurance; and retirement benefits, usually paid by the employer.

New nuclear engineers usually work under the supervision of experienced engineers. In large companies, new engineers may receive formal training in classrooms or seminars. As beginning engineers gain knowledge and experience, they move on to more difficult projects with greater independence to develop designs, solve problems, and make decisions.

Eventually, nuclear engineers may advance to become technical specialists or to supervise a team of engineers and technicians. Some may become engineering managers or move into sales work. For more information, see the profiles on architectural and engineering managers and sales engineers.

Nuclear engineers also can become medical physicists. A master's degree in health physics, radiological sciences, or a related field is necessary for someone to enter this field.

EMPLOYMENT AND OUTLOOK

There were approximately 17,200 nuclear engineers employed nationally in 2020. Employment of nuclear engineers is projected to decline 8 percent from 2020 to 2030.

Despite declining employment, about 900 openings for nuclear engineers are projected each year, on average, over the decade. All of those openings are expected to result from the need to replace workers who transfer to other occupations or exit the labor force, such as to retire.

Traditionally, utilities that own or build nuclear power plants have employed the greatest number of nuclear engineers. However, utilities often are opting for cheaper natural gas in power generation. In addition, the increasing viability of renewable energy is putting economic pressure on traditional nuclear power generation.

Related Occupations

- Civil Engineer
- Electrical/Electronic Engineering Technologist/Technician
- Electrical/Electronics Engineer
- Health & Safety Engineer
- Mechanical Engineer
- Physicist/Astronomer

MORE INFORMATION

American Nuclear Society (ANS)
555 N. Kensington Avenue
La Grange Park, Illinois 60526
800.323.3044
askanything@ans.org
www.ans.org

American Society for Engineering Education (ASEE)
1818 N Street, NW, Suite 600
Washington, DC 20036
202.331.3500
www.asee.org

Engineering Education Service Center (EESC)
1411 Old Hardman Btms Road
Clarkesville, GA 30523
706.499.5011
www.engineeringedu.com

Health Physics Society (HPS)
950 Herndon Parkway, Suite 450
Herndon, VA 20170
703.790.1745
hps@burkinc.com
hps.org

National Council of Examiners for Engineering and Surveying (NCEES)
200 Verdae Boulevard
Greenville, SC 29607
800.250.3196
www.ncees.org

National Society of Professional Engineers (NSPE)
1420 King Street
Alexandria, VA 22314
888.285.6773
www.nspe.org

Nuclear Energy Institute (NEI)
1201 F Street, NW, Suite 1100
Washington, DC 20004-1218
202.739.8000
www.nei.org

Society of Nuclear Medicine & Molecular Imaging (SNMMI)
1850 Samuel Morse Drive
Reston, VA 20190
703.708.9000
www.snmmi.org

Technology Student Association (TSA)
1904 Association Drive
Reston, VA 20191-1540
703.860.9000
general@tsaweb.org
tsaweb.org

Stuart Paterson

Conversation With...
SAM BRINTON

Senior Policy Analyst
Bipartisan Policy Center, Washington, D.C.
Nuclear engineer, 1 year

What was your individual career path in terms of education/training, entry-level job, or other significant opportunity?

I was a missionary kid and traveled all the time. I got to live in the Amazon jungle, but I also lived in dirty cities that were essentially concrete jungles. The pollution affected my allergies and I realized that clean air was really important. I thought one way to improve air quality would be to improve nuclear energy as an engineer. My mom had grown up near the Three Mile Island nuclear accident, so my passion for nuclear engineering kind of confused my family! As an undergrad, I studied nuclear engineering as well as music at Kansas State University. I went to Massachusetts Institute of Technology (MIT) for graduate school because I could study both technology and policy there. After that, I did a fellowship at Third Way, which is a think tank here in DC.

In my job as a Senior Policy Analyst, I'm kind of a translator, a technical translator. I'm not working in engineering, but because I have the technical background, I am respected.

My involvement in two groups, Nuclear Pride (for nuclear engineers who are part of the LGBT community) and the American Nuclear Society were critical for developing a network. From them, I acquired "soft skills" and learned who to turn to when I don't have the best answer to a question asked by a congressman, for instance.

What are the most important skills and/or qualities for someone in your profession?

A critical skill is a passion for systems. We do a massive amount of word problems based on how things work and are integrated with one another. Nuclear reactors run on physics, but you also need to know mechanical engineering, chemistry, math. The ability to wrap them all into one system is key. Even in my policy work, I need to bring together different parts of the system—this policy person, this business person, that senator—and integrate them to work together.

What do you wish you had known going into this profession?

I wish I had realized that science is always moving forward and the answers in your textbooks will change. Nuclear engineering involves fusion. It's on the very edge of what we know is possible in physics and engineering. You have to be able to go with the flow and not fight it.

Also, diversity might be a challenge. You won't always have professors or classmates who look like you or understand your perspective. I thought I was alone in this world, but Nuclear Pride made me realize I'm surrounded by members of the LGBT community.

Are there many job opportunities in your profession? In what specific areas?

There are growing opportunities in nuclear power, but it's a small industry. We're building four or five nuclear plants right now, each of which will be hiring nuclear engineers. There are dozens of nuclear startups. There's the entire field of nuclear waste. The joke is there's unlimited opportunity because the waste will be around for hundreds or thousands of years. You can also work in policy, as I do. The 20 to 30 nuclear engineering programs around the country offer the possibility of doing research or teaching at a university. You could also work for the Department of Energy or the Nuclear Regulatory Agency doing technology translation and helping people understand complicated topics.

How do you see your profession changing in the next five years? What role will technology play in those changes, and what skills will be required?

My hope is nuclear energy is going to be more compatible with the renewable community so that we're all working together. Nuclear has a very sordid and dirty history to some, but I think we can overcome that. Nuclear reactors are clean, safe and reliable. I think that with the next generation of reactors, we can be even better at safety and innovation.

What do you enjoy most about your job? What do you enjoy least about your job?

What I enjoy most is the flexibility and creativity. A nuclear engineer is not necessarily tackling the same problem over and over. I get to see new ideas and new conversations every week. And nuclear engineering is respectful of its next generation of engineers. Even a kid with a bright red mohawk like myself is listened to, because it's all about thinking outside the box and viewing a system from a variety of lenses.

What I like least is that you can't live everywhere you might want to. Working at a nuclear power plant, you'll probably be away from a city. If you're at a National Lab, you'll be even farther from any major metropolis.

Can you suggest a valuable "try this" for students considering a career in your profession?

Most nuclear power plants have a great tour where they tell you all about being a nuclear engineer. There are about 100 nuclear power plants across the country, so there may be one near you.

This interview was originally published in 2015.

Petroleum Engineer

Snapshot

Career Cluster(s): Science, Technology, Engineering & Mathematics

Interests: Oil and Gas; Earth Science; Geology

Earnings (Yearly Average): $137,330

Employment & Outlook: As Fast As Average Growth Expected

OVERVIEW

Sphere of Work

Petroleum engineers design and develop methods for extracting oil and gas from deposits below the Earth's surface. Petroleum engineers also find new ways to extract oil and gas from older wells.

Work Environment

Petroleum engineers typically work in offices, but are frequently required to travel to drill sites to inspect operations and to meet with colleagues and on-site workers.

Occupation Interest

Petroleum engineers have an interest not just in engineering science, but geology and the extraction of vital fossil fuels from the Earth. They are academically minded and want the opportunity to apply their knowledge in practical settings around the world.

Photo via iStock/industryview. [Used under license.]

A Day in the Life—Duties and Responsibilities

Oil and gas deposits, or reservoirs, are located deep in rock formations underground. These reservoirs can be accessed only by drilling wells, either on land, or at sea from offshore oil rigs.

Once oil and gas are discovered, petroleum engineers work with geoscientists and other specialists to understand the geologic formation of the rock containing the reservoir. They then determine the drilling methods, design the drilling equipment, implement the drilling plan, and monitor operations.

The best techniques currently being used recover only a portion of the oil and gas in a reservoir, so petroleum engineers also research and develop new ways to recover more of the oil and gas. This additional recovery helps to lower the cost of drilling and production.

Profile

Working Conditions: Both Inside and Outside
Physical Strength: Medium to Heavy Work
Education Needs: Bachelor's Degree; Master's Degree
Licensure/Certification: Varies
Opportunities for Experience: Co-op
Interest Score: IRE

OCCUPATION SPECIALTIES

Duties and Responsibilities

- Designing equipment to extract oil and gas from onshore and offshore reserves deep underground
- Developing plans to drill in oil and gas fields, and then to recover the oil and gas
- Developing ways to inject water, chemicals, gases, or steam into an oil reserve to force out more oil or gas
- Making sure that oilfield equipment is installed, operated, and maintained properly
- Evaluating the production of wells through surveys, testing, and analysis

Completions Engineer

Completions engineers decide the best way to finish building wells so that oil or gas will flow up from underground. They oversee work to complete the building of wells—a project that might involve the use of tubing, hydraulic fracturing, or pressure-control techniques.

Drilling Engineer

Drilling engineers determine the best way to drill oil or gas wells, taking into account a number of factors, including cost. They also ensure that the drilling process is safe, efficient, and minimally disruptive to the environment.

Production Engineer

Production engineers take over wells after drilling is completed. They typically monitor wells' oil and gas production. If wells are not producing as much as expected, production engineers figure out ways to increase the amount being extracted.

Reservoir Engineer

Reservoir engineers estimate how much oil or gas can be recovered from underground deposits, known as reservoirs. They study reservoirs' characteristics and determine which methods will get the most oil or gas out of the reservoirs. They also monitor operations to ensure that optimal levels of these resources are being recovered.

WORK ENVIRONMENT

Immediate Physical Environment

Petroleum engineers generally work in offices or at drilling and well sites. Travel is frequently required to visit these sites or to meet with other engineers, oilfield workers, and customers.

Large oil and gas companies maintain operations around the world; therefore, petroleum engineers sometimes work in other countries. Petroleum engineers also must be able to work with people from a wide variety of backgrounds, including other types of engineers, scientists, and oil and gas field workers.

Petroleum engineers typically work full time. Overtime may be necessary when traveling to and from drilling sites to help in their operation or respond to problems when they arise.

Human Environment

Petroleum engineers frequently work with other scientists and engineers such as geoscientists in the planning and execution of drill sites. They may need to travel frequently to these sites to inspect operations, and in so doing will meet with site supervisors and other professionals in the field, often in foreign countries and in potentially harsh conditions. Depending on whether they work as consultants or for an engineering firm, they may or may not report to superiors.

Technological Environment

Petroleum engineers commonly use development environment software and object or component-oriented development software, as well as analytical or scientific software, databases, and other typical office applications. They should be familiar with the drilling equipment and related tools and infrastructure used in the field, to better understand the complexities involved, and to aid in problem-solving and creative design for oil and gas extraction.

EDUCATION, TRAINING, AND ADVANCEMENT

High School/Secondary

High school students interested in studying petroleum engineering may benefit from taking classes in math, such as algebra, trigonometry, and calculus; and science, such as biology, chemistry, and physics.

Suggested High School Subjects

- Algebra
- Biology
- Calculus
- Chemistry
- Earth or Life or Physical Science
- Economics
- Entrepreneurship
- Geometry
- History
- Physics
- Pre-Calculus
- Statistics
- Trigonometry

Related Career Pathways/Majors

Science, Technology, Engineering and Mathematics Career Cluster

- Engineering & Technology Pathway
- Science & Mathematics Pathway

Transferable Skills and Abilities

Analytical Skills

- Compiling and making sense of large amounts of technical information and data in order to ensure that facilities operate safely and effectively

Creativity

- Coming up with creative designs to extract oil and gas, given the unique challenges posed by oil and gas sites in different regions of the world

Interpersonal Skills

- Working with others on projects that require highly complex machinery, equipment, and infrastructure
- Communicating and working well with other engineers and oil and gas workers to ensure that projects meet customer needs and run safely and efficiently

Math Skills

- Using the principles of calculus and other advanced topics in math for analysis, design, and troubleshooting in their work

Problem-solving Skills

- Identifying problems in drilling plans, since these problems can be costly
- Being careful not to overlook any potential issues and quickly addressing those which do occur

Postsecondary

College students typically pursue a bachelor's degree in engineering. Bachelor's degree engineering programs typically take four years and include academic, laboratory, and fieldwork in areas such as engineering principles, geology, and thermodynamics. Some colleges and universities offer cooperative programs in which students gain practical experience while completing their education.

Colleges and universities may offer five-year engineering programs that lead to both a bachelor's degree and a master's degree. Some employers prefer to hire candidates who have a master's degree. A graduate degree also allows an engineer to work as an instructor at some universities or in research and development (R&D).

Employers may prefer candidates who completed their studies in an engineering program accredited by a professional association such as the Accreditation Board for Engineering and Technology, Inc. (ABET).

Related College Majors

- Chemical Engineering
- Civil Engineering
- Environmental Engineering
- Geochemistry
- Geological Engineering
- Geology & Earth Sciences
- Geoscience Engineering
- Geophysics & Seismology
- Hydrology & Water Resources Science
- Mechanical Engineering
- Ocean Engineering
- Oceanography

Adult Job Seekers

Adults interested in becoming petroleum engineers should have a background in a related area such as geology, or engineering specialties such as civil, chemical, environmental, or mechanical engineering, and be prepared to transition their careers. They might be able to gain practical experience by taking on petroleum-related projects in their current jobs, if possible. Further coursework will likely also be required.

Professional Certification and Licensure

Licensure is not required for entry-level positions as a petroleum engineer. A Professional Engineering (PE) license, which allows for higher levels of leadership and independence, can be acquired later in one's career. Licensed engineers are called professional engineers (PEs). A PE can oversee the work of other engineers, sign off on projects, and provide services directly to the public. State licensure generally requires:

- A degree from an ABET-accredited engineering program
- A passing score on the Fundamentals of Engineering (FE) exam
- Relevant work experience, typically at least four years
- A passing score on the Professional Engineering (PE) exam

The initial FE exam can be taken after one earns a bachelor's degree. Engineers who pass this exam are commonly called engineers in training (EITs) or engineer interns (EIs). After meeting work experience requirements, EITs and EIs can take the second exam, called the Principles and Practice of Engineering (PE).

Several states require engineers to take continuing education courses in order to keep their licenses. Most states recognize licensure from other states if the licensing state's requirements meet or exceed their own licensure requirements. The Society of Petroleum Engineers offers certification. To be certified, petroleum engineers must be members of the Society, pass an exam, and meet other qualifications.

EARNINGS AND ADVANCEMENT

The median annual wage for petroleum engineers in 2020 was $137,330. The lowest 10 percent earned less than $78,620, and the highest 10 percent earned more than $208,000.

Petroleum engineers may receive benefits such as paid vacations, holidays, and sick days; life and health insurance; and retirement benefits, usually paid by the employer. They may also be entitled to recoup expenses related to work-related travel, or have an expense account for that purpose.

Entry-level engineers usually work under the supervision of experienced engineers. In large companies, new engineers also may receive formal training. As engineers gain knowledge and experience, they move to more difficult projects

Fast Fact

Petroleum products are not confined to the fuel industry—they are also used in the pharmaceutical and medical industries; for synthetic fabrics in the textile industry; to manufacture plastics in camping gear, sports gear, toys, and electronics; in the building of ships, trains, and plane parts; and in the synthesis of fertilizers, bubble gum, and perfumes. Source: drillers.com

Photo via iStock/IKvyatkovskaya. [Used under license.]

on which they have greater independence to develop designs, solve problems, and make decisions.

Eventually, petroleum engineers may advance to supervise a team of engineers and technicians. Some become engineering managers or move into other managerial positions. For more information, see the profile on architectural and engineering managers.

Petroleum engineers also may go into sales and use their engineering background to inform the discussion of a product's technical aspects with potential buyers and to help in product planning, installation, and use.

EMPLOYMENT AND OUTLOOK

There were approximately 28,500 petroleum engineers employed nationally in 2020. Employment is projected to grow 8 percent from 2020 to 2030, about as fast as the average for all occupations.

Oil prices will be a major determinant of employment growth. Because many petroleum engineers work in oil and gas extraction, any changes in oil prices will likely affect employment levels. Higher prices can cause oil and gas companies to increase capital investment in new facilities and expand existing production operations. Typically, companies also expand exploration for new reserves of oil and gas when prices are high.

Demand for petroleum engineers in support activities for mining should continue to be strong, as large oil and gas companies find it convenient and cost effective to contract production and drilling work to these firms as needed.

Related Occupations

- Architectural/Engineering Manager
- Chemical Engineer
- Civil Engineer
- Geological/Hydrologic Technician
- Geoscientist
- Industrial Engineer
- Mechanical Engineer
- Mining/Geological Engineer

MORE INFORMATION

Accreditation Board for Engineering and Technology, Inc. (ABET)
415 N. Charles Street
Baltimore, MD 21201
410.347.7700
www.abet.org

American Society for Engineering Education (ASEE)
1818 N Street, NW, Suite 600
Washington, DC 20036
202.331.3500
www.asee.org

National Council of Examiners for Engineering and Surveying (NCEES)
200 Verdae Boulevard
Greenville, SC 29607
800.250.3196
www.ncees.org

National Society of Professional Engineers (NSPE)
1420 King Street
Alexandria, VA 22314
888.285.6773
www.nspe.org

Society of Petroleum Engineers (SPE)
222 Palisades Creek Drive
Richardson, Texas 75080
972.952.9393
spedal@spe.org
www.spe.org

Technology Student Association (TSA)
1904 Association Drive
Reston, VA 20191-1540
703.860.9000
general@tsaweb.org
tsaweb.org

Stuart Paterson

Conversation With...
CHRISTA LAWSON

Subsea Operations Manager, for major oil company
Petroleum engineer, 19 years

What was your individual career path in terms of education/training, entry-level job, or other significant opportunity?

I have a B.S. in Mechanical Engineering. I began my career in oil and gas through an internship with a major oil company. That first summer, I spent many weeks going to offshore oil platforms that the company operated in the Gulf of Mexico. I got a great understanding of how oil and gas are produced and the incredible engineering that goes into bringing oil/gas from deep in the seabed, processing it and then transporting it onshore for refining.

I spent two summers interning with the company that I accepted a full time position with upon graduation. In my current role, I manage a team of subsea engineers responsible for the maintenance and repair of equipment, pipelines and structures siting on the ocean floor that are used to control and transport oil and gas.

What are the most important skills and/or qualities for someone in your profession?

Important skills for any engineer are logical and inquisitive thinking. You must be curious about how things work and want to problem solve. Also key to success is being a strong team player. I have seldom worked alone in this industry. Most work is done in dynamic teams, with each member bringing unique knowledge to the project.

Writing skills are very valuable. Early in my career, I wrote operating procedures manuals for use by offshore technicians in their day-to-day operation of equipment. Turning engineering knowledge into easy-to-understand-and-follow instructions ensures a safe and long running operation.

Lastly, you need initiative. You must be willing to stretch yourself and get outside your comfort zone to work on assignments that are unfamiliar. This industry has many challenges. It needs bright, ambitious engineers to secure its future.

What do you wish you had known going into this profession?

How much relationship building is required. I negotiate daily with my team and other teams. My ability to interface and to influence can affect the overall outcome. While these "soft skills" are necessary for success, they're not exactly cultivated in engineering school. Networking is important too. Take every opportunity to meet the leaders in your field. Talk to them, ask questions and don't be afraid to ask for assistance. My mentors have supported me and provided opportunities that took me further than I could ever have gone on my own.

Are there many job opportunities in your profession? In what specific areas?

The oil industry is largely driven by the price of a barrel of oil. When prices are high, there is ample opportunity for engineers of all backgrounds. When prices are low, opportunity shrinks. Some of the most interesting and well-paying jobs are in the "upstream"—or exploration and production—part of the petroleum sector.

How do you see your profession changing in the next five years? What role will technology play, and what skills will be required?

Technology is at the fore of all that we do. We are constantly trying to find new, safer, and environmentally conscious ways to produce oil and gas.

What do you enjoy most about your job? What do you enjoy least about your job?

My job is exciting. I get to travel on helicopters and large construction boats. I direct Remotely Operated Vehicles (ROV) working in water depths of over a mile. Rapidly changing technology and engineering advances mean there's always something new to learn. I like knowing that I help the world run. Energy and petroleum products— medical devices, plastic, fuel to heat homes and, of course, gasoline—are such an important part of everyone's day. In a small way, I know I am helping improve and make people's lives more comfortable.

Travel has been one of the most memorable parts of my work. I have travelled to Japan, the U.K., and many other interesting places. I even had a chance to live and work in Norway for two years as a part of an assignment.

Working in the petroleum industry can be exhausting. I don't always work 9 to 5. In the operations department, we're producing oil/gas 24 hours a day, seven days a week, including holidays. There are times you will need to go the extra mile, stay late, and even miss being with your family. I have spent several Thanksgivings on an offshore platform, but really I didn't mind because the cooks made an incredible feast!

Can you suggest a valuable "try this" for students considering a career in your profession?

Find the LinkedIn profile of someone working in the sector and email them for information. Also, oil services companies and major oil companies have cutting edge websites that give a glimpse into what they do.

By far the best "try this" would be to intern with an oil company. I knew after my first summer as an intern that this would be my lifelong career. It has been a great decision.

This interview was originally published in 2015.

Postsecondary Teacher

Snapshot

Career Cluster(s): Education & Training; Science, Technology, Engineering & Mathematics

Interests: Science; Teaching; Research

Earnings (Yearly Average): $80,560

Employment & Outlook: Faster Than Average Growth Expected

OVERVIEW

Sphere of Work

Postsecondary teachers instruct students in a variety of academic subjects beyond the high school level, including in engineering and other STEM-related areas. They may also conduct research and publish scholarly papers in peer-reviewed journals, and books.

Work Environment

Postsecondary teachers primarily work indoors, often in specialized classrooms and laboratories equipped with industry-specific technology for demonstrating tools and techniques to students. Depending on their area of academic study, they may undertake research projects either on their own, or with the help of some of their students. They may also teach part time, while working a separate job in their profession outside of school hours.

Photo via iStock/skynesher. [Used under license.]

Occupation Interest

Postsecondary teachers find it rewarding to impart their knowledge and expertise to adult students, and to use their position at a college or university to facilitate research. They have a lifelong passion for their subject and have usually undertaken a career in the field leading up to their teaching position, whether involving research, laboratory work, or fieldwork.

Profile

Working Conditions: Inside
Physical Strength: Medium Work
Education Needs: Master's Degree; Doctoral Degree
Licensure/Certification: Varies
Interest Score: SAI

A Day in the Life—Duties and Responsibilities

Postsecondary teachers, often referred to as professors or faculty, specialize in a variety of subjects and fields. At colleges and universities, professors are organized into departments by degree field, such as history, science, or business. A professor may teach one or more courses within that department. For example, a science professor may teach general principles of engineering, as well as a more specialized area such as aerospace engineering.

The duties of postsecondary teachers vary, often based on the size of their employing institution. In large colleges or universities, they may teach courses, conduct research or experiments, publish original research, apply for grants to fund their research, or supervise graduate teaching assistants. In small colleges and universities or in community colleges, they may spend most of their time teaching courses and working with students.

Duties and Responsibilities

- Developing an instructional plan (known as a course outline or syllabus) for the course(s) they teach and ensuring that it meets college and department standards
- Planning lessons and assignments
- Teaching courses in their subject area
- Assessing students' progress by grading assignments, papers, exams, and other work
- Advising students on courses to take and how to achieve their goals
- Working with colleagues to develop or modify the curriculum for a degree or certificate program involving in-person, online, or hybrid delivery of course material
- Staying informed about changes and innovations in their field
- Serving on academic or administrative committees, as needed

Full-time professors, particularly those who have tenure (that is, their position cannot be terminated without just cause), often are expected to devote a great deal of time on original research. Tenured professors must document their analyses or critical reviews and publish their research findings. They also may be required to serve on college and university committees.

Part-time professors, often known as adjunct professors, usually spend most of their time teaching students.

Professors may teach large classes of several hundred students (often with the help of graduate teaching assistants), smaller classes of about 40 to 50 students, seminars with a few students, or laboratories in which students practice the subject matter. Some teach online, either exclusively or in addition to providing live instruction.

Professors' tasks also may include collaborating with their colleagues and attending conferences to keep up with developments in their field.

WORK ENVIRONMENT

Immediate Physical Environment

Most postsecondary teachers work full time, although part-time work is common. Postsecondary teachers who work part time may offer instruction at several colleges or universities. Some have a full-time job in their field of expertise in addition to a part-time teaching position. For example, an active engineer may teach a course at night in addition to their full-time practical duties.

College and university courses are generally scheduled during the day, although some are offered in the evenings or on weekends to accommodate students who have jobs or other obligations.

Academic calendars typically include breaks, such as between terms. The availability and type of course offerings during the summer vary by institution. Although some postsecondary teachers provide instruction in summer courses, others use the time to conduct research or engage in professional development.

Postsecondary teachers' schedules generally are flexible. Full-time teachers typically need to be on campus to teach classes and have office hours but otherwise are free to set their own schedules.

Human Environment

Professors may teach large classes of several hundred students (often with the help of graduate teaching assistants), smaller classes of about 40 to 50 students, seminars with a few students, or laboratories in which students practice the subject matter. Professors' tasks also may include collaborating with their colleagues and attending conferences to keep up with developments in their field.

Technological Environment

Some professors teach online, either exclusively or in addition to providing live instruction, so must be familiar with all the tools and techniques needed to execute those lessons successfully and smoothly for students. They must also be aware of the latest advances in their field to better instruct their students.

EDUCATION, TRAINING, AND ADVANCEMENT

High School/Secondary

High school students with an interest in becoming teachers should focus on courses such as English that instill strong communication skills and should maintain high grades in all subjects. Students interested in teaching a single subject should become proficient in that subject—in the case of engineering, students should take STEM courses, and plan to pursue a branch of engineering in a postsecondary setting as well.

Suggested High School Subjects

- Algebra
- Biology
- Chemistry
- Civics
- Earth or Life or Physical Science
- Economics
- English
- Geometry
- History
- Physics
- Political Science
- Psychology
- Statistics

Related Career Pathways/Majors

Education & Training Career Cluster

- Teaching/Training Pathway

Science, Technology, Engineering & Mathematics Career Cluster

- Science & Mathematics Pathway

Transferable Skills and Abilities

Critical-thinking Skills

- Conducting original research and designing experiments
- Analyzing information logically

Interpersonal Skills

- Working well with others for tasks such as instructing students and serving on committees

Resourcefulness

- Presenting information in a way that students will understand
- Adapting to the different learning styles of students and being able to use technology for lessons or assignments

Speaking Skills

- Presenting lectures and providing feedback to students

Writing Skills

- Possessing strong writing ability to publish original research and analysis

Postsecondary

Postsecondary teachers who work for 4-year colleges and universities typically require a PhD or other doctoral degree in their field. For some specialties or for part-time positions, schools may hire those with a master's degree or who are doctoral degree candidates.

Doctoral programs usually take several years to complete, and students typically need a bachelor's or master's degree to enroll. Most PhD programs require students to author a doctoral dissertation, a paper presenting original research in their field of study, which they then defend in questioning from experts. Candidates usually specialize in a subfield, such as organic chemistry or European history.

Community colleges may hire those with a master's degree. However, some institutions prefer that applicants have a PhD.

Related College Majors

- Education
- Engineering
- English/Liberal Arts
- Environmental Studies
- Geological Sciences
- Mathematics
- Physical Sciences
- STEM

Adult Job Seekers

It is quite common for active scientists to also pursue a career in postsecondary teaching. With greater experience and expertise comes greater opportunities in the teaching field, and some may choose to segue from part-time to full-time teaching and research at their hiring university.

Professional Certification and Licensure

Postsecondary teachers who prepare students for an occupation that requires a license, certification, or registration, may need to have—or may benefit from having—the same credential. For example, a postsecondary nursing teacher might need a nursing license, or a postsecondary education teacher might need a teaching license.

Fast Fact

In the U.S., teachers are often considered second only to military personal as the occupation that contributes most to the well-being of society. Source: www.factretriever.com

Photo via iStock/skynesher. [Used under license.]

Additional Requirements

Institutions may prefer to hire those with teaching or other work experience. In some fields, such as health specialties, art, law, and education, hands-on work experience is especially important. Postsecondary teachers in these fields often gain experience by working in an occupation related to their field of study.

In other fields, such as biological science, physics, and chemistry, some postsecondary teachers have postdoctoral research experience. Sometimes called a "postdoc," this experience takes the form of a job that usually involves working for 2 to 3 years as a research associate or in a similar position, often at a college or university.

Some postsecondary teachers gain teaching experience by working as graduate teaching assistants—students who are enrolled in a graduate program and teach classes at the institution in which they are enrolled.

EARNINGS AND ADVANCEMENT

Earnings of postsecondary teachers depend on experience, subjects taught, school, and location. Median annual earnings for postsecondary teachers were $80,560 in 2020 ($103,600 for engineering specifically). The lowest 10 percent earned less than $40,830, and the highest 10 percent earned more than $179,540.

Postsecondary teachers may receive paid vacations, holidays, and sick days; life and health insurance; and retirement benefits. These are usually paid by the employer.

Postsecondary teachers with a doctoral degree often seek tenure—a guarantee that a professor cannot be dismissed without just cause. Attaining tenure may take up to 7 years of progressing through the positions by rank: assistant professor, associate professor, and professor. The decision to grant tenure is based on a candidate's published research, contribution to the institution, and teaching.

Some professors advance to high-level administrative positions, such as dean or president. For information on deans and other administrative positions, see the profile on postsecondary education administrators.

EMPLOYMENT AND OUTLOOK

Postsecondary teachers held 1,276,900 jobs in 2020. Of those, 46,300 were employed in biological sciences. Overall employment of postsecondary teachers is projected to grow 12 percent from 2020 to 2030, faster than the average for all occupations. Both part-time and full-time postsecondary teachers are included in this projection.

The number of people attending postsecondary institutions is expected to grow over the projected decade. Students will continue to seek higher education to gain the knowledge and skills necessary to meet their career goals. As more people enter colleges and universities, more postsecondary teachers will be needed to serve these additional students. Colleges and universities are likely to hire more part-time teachers to meet this demand. In all disciplines, there is expected to be a limited number of full-time nontenure and full-time tenure positions.

As an aging population increasingly demands healthcare services, additional postsecondary teachers are expected to be needed to help educate the workers who provide these services.

However, despite expected increases in enrollment, employment growth in public colleges and universities will depend on state and local government budgets. If budgets for higher education are reduced, employment growth may be limited.

About 139,600 openings for postsecondary teachers are projected each year, on average, over the decade. Many of those openings are expected to result from the need to replace workers who transfer to different occupations or exit the labor force, such as to retire.

Related Occupations

- Career/Technical Education Teacher
- Elementary/Middle/High School Principal
- Instructional Coordinator
- Postsecondary Education Administrator

MORE INFORMATION

American Association of Colleges for Teacher Education (AACTE)
1602 L Street NW, Suite 601
Washington, DC 20036
202.293.2450
aacte.org

American Association of University Professors (AAUP)
1133 Nineteenth Street NW, Suite 200
Washington, DC 20036
202.737.5900
communications@aaup.org
www.aaup.org

American Educational Research Association (AERA)
1430 K Street NW, Suite 1200
Washington, DC 20005
202.238.3200
www.aera.net

Association of American Colleges and Universities (AACU)
1818 R Street NW
Washington, DC 20009
202.387.3760
information@aacu.org
www.aacu.org

Association of American Educators (AAE)
25909 Pala Place, Suite 330
Mission Viejo, CA 92691
800.704.7799
www.aaeteachers.org

Association of College and University Educators (ACUE)
745 Fifth Avenue, Suite 500
New York, NY 10151
866.628.2283
acue.org

Association of Teacher Educators (ATE)
11350 Random Hills Road, Suite 800, PMB 6
Fairfax, VA 22030
703.659.1708
info@ate1.org
www.ate1.org

Council of Graduate Schools (CGS)
One Dupont Circle NW, Suite 230
Washington, DC 20036
202.223.3791
general_inquiries@cgs.nche.edu
cgsnet.org

National Association of Biology Teachers (NABT)
P.O. Box 3363
Warrenton, VA 20188
703.264.9696
office@nabt.org
nabt.org

Stuart Paterson

Sales Engineer

Snapshot

Career Cluster(s): Human Services; Marketing; Science, Technology, Engineering & Mathematics

Interests: Engineering; Sales; Technological Products

Earnings (Yearly Average): $108,830

Employment & Outlook: As Fast As Average Growth Expected

OVERVIEW

Sphere of Work

Sales engineers sell complex scientific and technological products or services to businesses. They must have extensive knowledge of the products' parts and functions and must understand the scientific processes that make these products work.

Work Environment

Sales engineers frequently travel as part of their job, and may be away from home for extended periods. Their job can be stressful because their income depends on earning commission from successful sales.

Occupation Interest

Sales engineers have a solid foundation in engineering science, but also have a natural aptitude for sales and marketing and seek to combine those two strengths. They work well under pressure and are people-oriented. They don't mind the extensive travel required to carry out their job, and may even look forward to going on the road.

Photo via iStock/RossHelen. [Used under license.]

A Day in the Life—Duties and Responsibilities

Profile

Working Conditions: Both Inside and Outside
Physical Strength: Light to Medium Work
Education Needs: Bachelor's Degree
Licensure/Certification: Not Required
Opportunities for Experience: Work Experience; On-the-Job Training
Interest Score: ERI

Sales engineers specialize in technologically and scientifically advanced products. They use their technical skills to explain the benefits of their products or services to potential customers and to show how their products or services are better than their competitors'. Some sales engineers work for the companies that design and build technical products. Others work for independent sales firms.

Many of the duties of sales engineers are similar to those of other salespersons. They must interest the client in buying their products or services, negotiate a price, and complete the sale. To do this, sales engineers give technical presentations during which they explain the technical aspects of the product and how it will solve a specific customer problem.

Duties and Responsibilities

- Preparing and delivering technical presentations explaining products or services to existing and prospective customers
- Talking with customers and engineers to assess equipment needs and to determine system requirements
- Collaborating with sales teams to understand customer requirements and provide sales support
- Securing and renewing orders and arranging delivery
- Planning and modifying products to meet customer needs
- Helping clients solve problems with installed equipment
- Recommending improved materials or machinery to customers, showing how changes will lower costs or increase production
- Helping in researching and developing new products

Some sales engineers team with other salespersons, such as wholesale and manufacturing sales representatives, who concentrate on marketing and selling the product, which lets the sales engineer concentrate on the technical aspects of the job. By working as part of a sales team, each member is able to focus on his or her strengths and expertise.

In addition to giving technical presentations, sales engineers are increasingly doing other tasks related to sales, such as market research. They also may ask for technical requirements from customers and modify and adjust products to meet customers' specific needs. Some sales engineers work with research and development (R&D) departments to help identify and develop new products.

WORK ENVIRONMENT

Immediate Physical Environment

Some sales engineers have large territories and travel extensively. Because sales regions may cover several states, sales engineers may be away from home for several days or even weeks at a time. Other sales engineers cover a smaller region and spend only a few nights away from home.

Sales engineers may encounter stress because their income and job security often depend directly on their success in sales and customer service.

Most sales engineers work full time. Some may work additional and irregular hours to meet sales goals and client needs.

Human Environment

Sales engineers interact with many different people throughout the course of their work. As part of an engineering firm, they may work on the projects themselves that they then showcase on the road. If they work for a sales firm, they need to meet with engineering clients in order to better understand the products that they will be selling. When on sales trips, they must meet with numerous clients and give technical presentations in order to demonstrate the merits of the product they are trying to sell. They must be able to handle rejection and maintain a positive attitude while working under considerable stress and prolonged absences from home. They may also work with other members of a sales team.

Technological Environment

Sales engineers must be familiar with engineering and technological practices in order to convincingly sell these products. Furthermore, they must understand their products well enough to answer questions that arise during sales meetings. They will utilize presentation software and other computer applications while making technical sales presentations to clients.

EDUCATION, TRAINING, AND ADVANCEMENT

High School/Secondary

Students interested in becoming sales engineers do not necessarily need to first become engineers, but it is strongly recommended in order to gain knowledge of technology and industry practices. Students should consider which field of engineering interests them and work towards that goal, excelling in math and science courses along the way.

Suggested High School Subjects

- Algebra
- Biology
- Calculus
- Chemistry
- Earth or Life or Physical Science
- Economics
- Entrepreneurship
- Geography
- Geometry
- Physics
- Pre-Calculus
- Psychology
- Statistics
- Trigonometry

Transferable Skills and Abilities

Interpersonal Skills

- Building relationships with clients and effectively communicating with other members of the sales team

Problem-solving Skills

- Listening to the customer's desires and concerns, and then recommend solutions, such as customizing a product for the customer

Self-confidence

- Confidence and persuasiveness when making sales presentations

Technological Skills

- Possessing extensive knowledge of the technologically sophisticated products they sell in order to explain their advantages and answer questions

Related Career Pathways/Majors

Marketing Career Cluster

- Professional Sales Pathway

Human Services Career Cluster

- Consumer Services Pathway

Science, Technology, Engineering & Mathematics Career Cluster

- Engineering & Technology Pathway
- Science & Mathematics Pathway

Postsecondary

Sales engineers typically need a bachelor's degree in engineering or a related field. University engineering programs generally require four years of study. They vary in content, but all programs include courses in math and the physical sciences. In addition, most programs require developing strong computer skills.

Most engineering programs require students to choose an area of specialization. The most common majors are electrical, mechanical, or civil engineering, but some engineering departments offer additional majors, such as chemical, biomedical, or computer hardware engineering. However, some undergraduate programs offer a general engineering curriculum; students then specialize in a particular area either on the job or in graduate school.

Related College Majors

- Biomedical Engineering
- Business
- Chemical Engineering
- Civil Engineering
- Communications
- Computer Hardware Engineering
- Electrical Engineering
- Engineering (General)
- Finance
- Marketing
- Mechanical Engineering
- Public Relations
- Technology

Adult Job Seekers

A worker without a degree, but with previous sales experience as well as technical experience or training, may become a sales engineer. Workers who have a degree in a science, such as chemistry, or in business with little or no previous sales experience, also may become sales engineers.

Additional Requirements

New graduates with engineering degrees typically need sales experience and training before they can work independently as sales engineers. Training covers general sales techniques and may involve teaming with a sales mentor who is familiar with the employer's business practices, customers, procedures, and company culture. After the training period, sales engineers may continue to partner with someone who lacks technical skills yet excels in the art of sales.

It is important for sales engineers to continue their engineering and sales education throughout their careers. Much of their value to their employers depends on their knowledge of, and ability to sell, the latest technologies. Sales engineers in high-technology fields, such as information technology and advanced electronics, may find that their technical knowledge rapidly becomes obsolete, requiring frequent retraining.

EARNINGS AND ADVANCEMENT

The median annual wage for sales engineers was $108,830 in 2020. The lowest 10 percent earned less than $60,730, and the highest 10 percent earned more than $184,190.

How much a sales engineer earns varies considerably by the type of firm and the product sold. Most employers offer a combination of salary and commission payments or salary plus a bonus. Some sales engineers who work for independent sales companies earn only commissions.

Commissions are usually based on the value of sales. Bonuses may depend on individual performance, on the performance of all workers in the group or district, or on the company's performance. Earnings from commissions and bonuses may vary from year to year depending on sales ability, the demand for the company's products or services, and the overall economy.

In addition to their earnings, sales engineers who work for manufacturers are usually reimbursed for expenses such as transportation, meals, hotels, and customer entertainment.

Promotions may include a higher commission rate, a larger sales territory, or elevation to the position of supervisor or sales manager.

EMPLOYMENT AND OUTLOOK

Employment of sales engineers is projected to grow 8 percent from 2020 to 2030, about as fast as the average for all occupations.

As technologically sophisticated products continue to reach the market, sales engineers will be in demand to help sell products or services related to these products.

Employment growth is likely to be strong for sales engineers selling computer software and hardware. Strong industry growth is expected in computer systems design and related services, contributing to employment growth in the occupation.

Sales engineers also are expected to see employment growth in independent sales agencies (companies that sell on behalf of manufacturers without taking title to the goods being sold). As manufacturing companies outsource their sales staff in an effort to control costs, employment in these independent agencies should increase.

Related Occupations

- Aerospace Engineer
- Computer Hardware Engineer
- Electrical/Electronics Engineer
- Industrial Engineer
- Insurance Sales Agent
- Materials Engineer
- Mechanical Engineer
- Purchasing Manager/Buyer/Purchasing Agent
- Securities/Commodities/Financial Services Sales Agent
- Wholesale/Manufacturing Sales Representative

MORE INFORMATION

Manufacturers' Agents National Association (MANA)
6321 W. Dempster Street, Suite 110
Morton Grove, IL 60053
877.626.2776
mana@manaonline.org
www.manaonline.org

Manufacturers' Representatives Educational Research Foundation (MRERF)
1301 Arapahoe Street, Suite 105
Golden, CO 80401
303.463.1801
certify@mrerf.org
mrerf.org

Stuart Paterson

Conversation With...
EVAN MULVIHILL

Sales Engineer
Sweetwater Sound, Inc., Fort Wayne, IN
Sales engineer, 3 years

What was your individual career path in terms of education/training, entry-level job, or other significant opportunity?

I went to school for music composition at the University of Illinois and for recording arts at Tribeca Flashpoint College, a media, arts, and communications school in Chicago. I'm a musician and play several instruments, but piano is my "forte." (Musician joke!) Throughout school, I gained experience by doing freelance engineering work, such as mixing and recording bands and post-production/location sound for video. I was a boom operator for several feature-length films and helped out with sound at a 40,000-square-foot haunted house called Fear City. When I was hired as a sales engineer by Sweetwater Sound, I had to complete a three-month training and certificate program known as "Sweetwater University" before rolling out onto the sales floor. I liken the program to drinking from a fire hose because the amount of information you're required to absorb is like getting a master's degree in thirteen weeks.

What are the most important skills and/or qualities for someone in your profession?

Because a sales engineer is essentially a salesman with expert knowledge in a given technical field, being knowledgeable in your field is obviously important, as is the willingness to learn new information and remain relevant. Being a problem-solver and being a people person are important too. You need to be compassionate when a customer calls with some issue they're struggling with. Even if you don't know the answer or can't solve their problem, if you show them you care, that's the start of a long-term relationship. A quote that's always resonated with me is "Nobody cares how much you know until they know how much you care." But if I had to choose one quality that's a must, it's the ability to strive to improve what you aren't good at and to refine what you are good at.

What do you wish you had known going into this profession?

As much as I love studying all the minutia of audio technology, what I've realized over the years is that technical details often don't matter to the customer and rarely, if ever, come up. What's more important is to aid the decision making-process and help them feel that they've made the right choice.

Are there many job opportunities in your profession? In what specific areas?

The music industry in general has very limited job opportunities, but it's a different story for the equipment-sales segment of the industry. Because of piracy, making it as a band or finding paid work in recording studios is almost impossible. However, because musicians are now buying equipment for touring or for producing their own music at home, the retail side of the music industry is booming. During the gold rush, a prospector could work all day and never strike gold; however, there was always business for the guy selling him the tools.

How do you see your profession changing in the next five years? What role will technology play in those changes, and what skills will be required?

That's hard to say. Music technology is constantly changing, with products being released and discontinued almost every month. It's hard to know what new products are being dreamt up right now. But given how rapidly technology evolves in this industry, all I can say is I'm excited to see what's in store. Of course, the skills that will be required are technological know-how and the willingness to learn how to use and master new products.

What do you enjoy most about your job? What do you enjoy least about your job?

I love learning about and having access to cool new products, but I really love the people side of the business, too. I love my customers, many of whom I've worked with for years and consider friends. I also love the people I work with, my co-workers and managers. I love having job flexibility. I feel fortunate to work for a company that goes above and beyond to make sure all of its employees are taken care of and happy.

Most of my customers are really awesome people, but what I like least are the small handful who aren't pleasant to talk to or that you have to tiptoe around because they'll blow up over the smallest thing.

Can you suggest a valuable "try this" for students considering a career in your profession?

Being a professional salesperson is all about figuring out what issues the customer is facing and fixing them. Sometimes the solution to their problem is a product you can sell them. Sometimes it's just advice or resources. Try this: The next time you strike up a conversation with someone, find out if they're having difficulty with something and help them fix it. Being successful at sales starts with a genuine interest in helping people.

This interview was originally published in 2016.

What Are Your Career Interests?

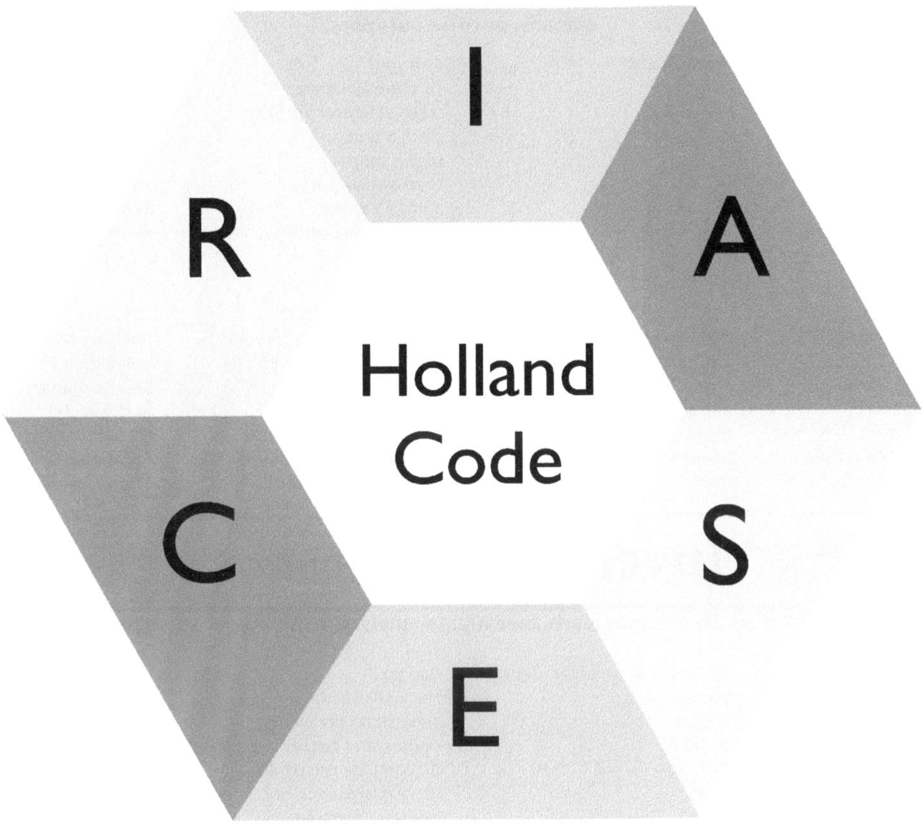

This is based on Dr. John Holland's theory that people and work environments can be loosely classified into six different groups. Each of the letters above corresponds to one of the six groups described in the following pages.

Different people's personalities may find different environments more to their liking. While you may have some interests in and similarities to several of the six groups, you may be attracted primarily to two or three of the areas. These two or three letters are your "Holland Code." For example, with a code of "RES" you would most resemble the Realistic type, somewhat less resemble the Enterprising type, and resemble the Social type even less. The types that are not in your code are the types you resemble least of all.

Most people, and most jobs, are best represented by some combination of two or three of the Holland interest areas. In addition, most people are most satisfied if there is some degree of fit between their personality and their work environment.

The rest of the pages in this booklet further explain each type and provide some examples of career possibilities, areas of study at MU, and co-curricular activities for each code. To take a more in-depth look at your Holland Code, take a self-assessment such as the SDS, Discover, or a card sort at the MU Career Center with a Career Specialist.

Realistic *(Doers)*

People who have athletic ability, prefer to work with objects, machines, tools, plants or animals, or to be outdoors.

Are you?
practical
straightforward/frank
mechanically inclined
stable
concrete
reserved
self-controlled
independent
ambitious
systematic

Can you?
fix electrical things
solve electrical problems
pitch a tent
play a sport
read a blueprint
plant a garden
operate tools and machine

Like to?
tinker with machines/vehicles
work outdoors
be physically active
use your hands
build things
tend/train animals
work on electronic equipment

Career Possibilities (Holland Code):

Air Traffic Controller (SER)
Archaeologist (IRE)
Athletic Trainer (SRE)
Cartographer (IRE)
Commercial Airline Pilot (RIE)
Commercial Drafter (IRE)
Corrections Officer (SER)
Dental Technician (REI)
Farm Manager (ESR)
Fish and Game Warden (RES)
Floral Designer (RAE)
Forester (RIS)
Geodetic Surveyor (IRE)
Industrial Arts Teacher (IER)
Laboratory Technician (RIE)
Landscape Architect (AIR)
Mechanical Engineer (RIS)
Optician (REI)
Petroleum Geologist (RIE)
Police Officer (SER)
Practical Nurse (SER)
Property Manager (ESR)
Recreation Manager (SER)
Service Manager (ERS)
Software Technician (RCI)
Ultrasound Technologist (RSI)
Vocational Rehabilitation Consultant (ESR)

Investigative *(Thinkers)*

People who like to observe, learn, investigate, analyze, evaluate, or solve problems.

Are you?
inquisitive
analytical
scientific
observant/precise
scholarly
cautious
intellectually self-confident
Independent
logical
complex
Curious

Can you?
think abstractly
solve math problems
understand scientific theories
do complex calculations
use a microscope or computer
interpret formulas

Like to?
explore a variety of ideas
work independently
perform lab experiments
deal with abstractions
do research
be challenged

Career Possibilities (Holland Code):

Actuary (ISE)
Agronomist (IRS)
Anesthesiologist (IRS)
Anthropologist (IRE)
Archaeologist (IRE)
Biochemist (IRS)
Biologist (ISR)
Chemical Engineer (IRE)
Chemist (IRE)
Computer Systems Analyst (IER)
Dentist (ISR)
Ecologist (IRE)
Economist (IAS)
Electrical Engineer (IRE)
Geologist (IRE)
Horticulturist (IRS)
Mathematician (IER)
Medical Technologist (ISA)
Meteorologist (IRS)
Nurse Practitioner (ISA)
Pharmacist (IES)
Physician, General Practice (ISE)
Psychologist (IES)
Research Analyst (IRC)
Statistician (IRE)
Surgeon (IRA)
Technical Writer (IRS)
Veterinarian (IRS)

Artistic *(Creators)*

People who have artistic, innovating, or intuitional abilities and like to work in unstructured situations using their imagination and creativity.

Are you?
creative
imaginative
innovative
unconventional
emotional
independent
Expressive
original
introspective
impulsive
sensitive
courageous
complicated
idealistic
nonconforming

Can you?
sketch, draw, paint
play a musical instrument
write stories, poetry, music
sing, act, dance
design fashions or interiors

Like to?
attend concerts, theatre, art exhibits
read fiction, plays, and poetry
work on crafts
take photography
express yourself creatively
deal with ambiguous ideas

Career Possibilities (Holland Code):

Actor (AES)
Advertising Art Director (AES)
Advertising Manager (ASE)
Architect (AIR)
Art Teacher (ASE)
Artist (ASI)
Copy Writer (ASI)
Dance Instructor (AER)
Drama Coach (ASE)
English Teacher (ASE)
Entertainer/Performer (AES)
Fashion Illustrator (ASR)
Interior Designer (AES)
Intelligence Research Specialist (AEI)
Journalist/Reporter (ASE)
Landscape Architect (AIR)
Librarian (SAI)
Medical Illustrator (AIE)
Museum Curator (AES)
Music Teacher (ASI)
Photographer (AES)
Writer (ASI)
Graphic Designer (AES)

Social *(Helpers)*

People who like to work with people to enlighten, inform, help, train, or cure them, or are skilled with words.

Are you?
friendly
helpful
idealistic
insightful
outgoing
understanding
cooperative
generous
responsible
forgiving
patient
kind

Can you?
teach/train others
express yourself clearly
lead a group discussion
mediate disputes
plan and supervise an activity
cooperate well with others

Like to?
work in groups
help people with problems
do volunteer work
work with young people
serve others

Career Possibilities (Holland Code):

City Manager (SEC)
Clinical Dietitian (SIE)
College/University Faculty (SEI)
Community Org. Director (SEA)
Consumer Affairs Director (SER)Counselor/Therapist (SAE)
Historian (SEI)
Hospital Administrator (SER)
Psychologist (SEI)
Insurance Claims Examiner (SIE)
Librarian (SAI)
Medical Assistant (SCR)
Minister/Priest/Rabbi (SAI)
Paralegal (SCE)
Park Naturalist (SEI)
Physical Therapist (SIE)
Police Officer (SER)
Probation and Parole Officer (SEC)
Real Estate Appraiser (SCE)
Recreation Director (SER)
Registered Nurse (SIA)
Teacher (SAE)
Social Worker (SEA)
Speech Pathologist (SAI)
Vocational-Rehab. Counselor (SEC)
Volunteer Services Director (SEC)

Enterprising *(Persuaders)*

People who like to work with people, influencing, persuading, leading or managing for organizational goals or economic gain.

Are you?
self-confident
assertive
persuasive
energetic
adventurous
popular
ambitious
agreeable
talkative
extroverted
spontaneous
optimistic

Can you?
initiate projects
convince people to do things your way
sell things
give talks or speeches
organize activities
lead a group
persuade others

Like to?
make decisions
be elected to office
start your own business
campaign politically
meet important people
have power or status

Career Possibilities (Holland Code):

Advertising Executive (ESA)
Advertising Sales Rep (ESR)
Banker/Financial Planner (ESR)
Branch Manager (ESA)
Business Manager (ESC)
Buyer (ESA)
Chamber of Commerce Exec (ESA)
Credit Analyst (EAS)
Customer Service Manager (ESA)
Education & Training Manager (EIS)
Emergency Medical Technician (ESI)
Entrepreneur (ESA)
Foreign Service Officer (ESA)
Funeral Director (ESR)
Insurance Manager (ESC)
Interpreter (ESA)
Lawyer/Attorney (ESA)
Lobbyist (ESA)
Office Manager (ESR)
Personnel Recruiter (ESR)
Politician (ESA)
Public Relations Rep (EAS)
Retail Store Manager (ESR)
Sales Manager (ESA)
Sales Representative (ERS)
Social Service Director (ESA)
Stockbroker (ESI)
Tax Accountant (ECS)

Conventional *(Organizers)*

People who like to work with data, have clerical or numerical ability, carry out tasks in detail, or follow through on others' instructions.

Are you?
well-organized
accurate
numerically inclined
methodical
conscientious
efficient
conforming
practical
thrifty
systematic
structured
polite
ambitious
obedient
persistent

Can you?
work well within a system
do a lot of paper work in a short time
keep accurate records
use a computer terminal
write effective business letters

Like to?
follow clearly defined procedures
use data processing equipment
work with numbers
type or take shorthand
be responsible for details
collect or organize things

Career Possibilities (Holland Code):

Abstractor (CSI)
Accountant (CSE)
Administrative Assistant (ESC)
Budget Analyst (CER)
Business Manager (ESC)
Business Programmer (CRI)
Business Teacher (CSE)
Catalog Librarian (CSE)
Claims Adjuster (SEC)
Computer Operator (CSR)
Congressional-District Aide (CES)
Cost Accountant (CES)
Court Reporter (CSE)
Credit Manager (ESC)
Customs Inspector (CEI)
Editorial Assistant (CSI)
Elementary School Teacher (SEC)
Financial Analyst (CSI)
Insurance Manager (ESC)
Insurance Underwriter (CSE)
Internal Auditor (ICR)
Kindergarten Teacher (ESC)
Medical Records Technician (CSE)
Museum Registrar (CSE)
Paralegal (SCE)
Safety Inspector (RCS)
Tax Accountant (ECS)
Tax Consultant (CES)
Travel Agent (ECS)

BIBLIOGRAPHY

Anderson, Ross. *Security Engineering: A Guide to Building Dependable Distributed Systems*, 3rd ed. Hoboken, NJ: Wiley, 2020.

Andrews, Gordon, Patricia Shaw and John McPhee. *Canadian Professional Engineering and Geoscience Practice and Ethics*, 6th ed. Toronto, Canada: Top Hat, 2018.

Ashby, Darren. *Electrical Engineering 101: Everything You Should Have Learned in School...but Probably Didn't*, 3rd ed. Amsterdam, Netherlands: Newnes, 2011.

Blockley, David. *Engineering: A Very Short Introduction.* Toronto, Canada: Oxford University Press, 2012.

Brain, Marshall. *The Engineering Book: From the Catapult to the Curiosity Rover, 250 Milestones in the History of Engineering.* New York: Sterling, 2015.

Brauer, Roger L. *Safety and Health for Engineers*, 3rd ed. Hoboken, NJ: Wiley, 2016.

Bunnell, Brian and Samer Najia. *Mechanical Engineering for Makers: A Hands-on Guide to Designing and Making Physical Things*. Santa Rosa, CA: Make Community, LLC, 2020.

Burkov, Andriy. *Machine Learning Engineering.* Quebec, Canada: True Positive Inc., 2020.

Cengel, Yunus and Michael Boles. *Thermodynamics: An Engineering Approach*, 8th ed. Toronto, Canada: McGraw-Hill Education, 2014.

Coduto, Donald, Man-chu Ronald Yeung and William Kitch. *Geotechnical Engineering: Principles and Practices*, 2nd ed. Toronto, Canada: Pearson, 2010.

Das, Braja J. *Principles of Geotechnical Engineering*, 10th ed. Boston, MA: Cengage Learning, 2020.

Ebeling, Charles E. *An Introduction to Reliability and Maintainability Engineering*, 3rd ed. Long Grove, IL: Waveland Press, 2019.

Enge, Per, Nick Enge and Stephen Zoepf. *Electric Vehicle Engineering.* Toronto, Canada: McGraw-Hill Education, 2020.

Engineers: From the Great Pyramids to the Pioneers of Space Travel. London: DK, 2015.

Fanchi, John R. and Richard L. Christiansen. *Introduction to Petroleum Engineering.* Hoboken, NJ: Wiley, 2016.

Fogler, H. Scott. *Elements of Chemical Reaction Engineering*, 5th ed. Hoboken, NJ: Prentice Hall, 2016.

Goetsch, David and Gene Ozon. *Occupational Health and Safety for Technologists, Engineers, and Managers*, 2nd ed. Toronto, Canada: Pearson Canada Inc, 2018.

Gordon, Henry A. *Mining and Engineering, and Miner's Guide.* France: Ulan Press, 2012.

Hamilton, Nathan. *Agricultural Engineering: Emerging Trends.* New York: Syrawood Publishing House, 2022.

Hamming, Richard R. *The Art of Doing Science and Engineering: Learning to Learn.* Boca Raton, FL: CRC Press, 1997.

Hartman, Howard L. and Jan M. Mutmansky. *Introductory Mining Engineering*, 2nd ed. Hoboken, NJ: Wiley, 2002.

How Things Work 2nd Edition: An Illustrated Guide to the Mechanics Behind the World Around Us. New York: Chartwell Books, 2019.

Howard, Albert. *An Agricultural Testament.* Hamburg, Germany: Albatross Publishers, 2018.

Katz, Bob. *Mastering Audio: The Art and the Science.* Oxfordshire, UK: Routledge, 2002.

Kemp, Albert W. *Industrial Mechanics*, 4th ed. Orland Park IL: American Tech Publishers, 2016.

Kuprenas, John and Matthew Fredrick. *101 Things I've Learned in Engineering School.* Victoria, Canada: Crown, 2018.

Mays, Larry W. *Water Resources Engineering*, 3rd ed. Hoboken, NJ: Wiley, 2019.

Mihelcic, James R. and Julie B. Zimmerman. *Environmental Engineering: Fundamentals, Sustainability, Design,* 3rd ed. Hoboken, NJ: Wiley, 2021.

Nellis, G.F. and S.A. Klein. *Introduction to Engineering Heat Transfer.* Cambridge, UK: Cambridge University Press, 2020.

Ostafichuk, Peter, Markus Fengler and Antony Hodgson. *The Engineering Design Process: An Introduction for Mechanical Engineers*, 3rd ed. Ottawa, Canada: Library and Archives Canada, 2018.

Petroski, Henry. *Design Paradigms: Case Histories of Error and Judgement in Engineering.* Cambridge, UK: Cambridge University Press, 1994.

———. *To Engineer Is Human: The Role of Failure in Successful Design.* United Kingdom: Vintage, 1992.

Plantenberg, Kirstie. *Engineering Graphics Essentials with AutoCAD 2022 Instruction: Text and Video Instruction.* Mission, KS: SDC Publications, 2021.

Russell, Paul A., Leslie Jackson and Thomas D. Morton. *Reeds Marine Engineering and Technology, Vol. 8: General Engineering Knowledge For Marine Engineers*, 6th ed. Auckland, NZ: Reeds, 2018.

Skakoon, James G. and W.J. King. *The Unwritten Laws of Engineering*. New York: ASME Press, 2019.

Winchester, Simon. *The Perfectionists: How Precision Engineers Created the Modern World.* New York: Harper Perennial, 2019.

Winters, Titus, Tom Manshrek and Hyrum Wright. *Software Engineering at Google: Lessons Learned from Programming Over Time.* Sebastopol, CA: O'Reilly Media, 2020.

LIST OF ORGANIZATIONS AND RESOURCES

Accreditation Board for Engineering and Technology, Inc. (ABET)
415 North Charles Street
Baltimore, MD 21201
410.347.7700
www.abet.org

Advance CTE: State Leaders Connecting Learning to Work
8484 Georgia Avenue, Suite 620
Silver Spring, MD 20910
301.588.9630
careertech.org

American Academy of Environmental Engineers and Scientists (AAEES)
147 Old Solomons Island Road, Suite 303
Annapolis, MD 21401
410.266.3311
info@aaees.org
www.aaees.org

American Association of Colleges for Teacher Education (AACTE)
1602 L Street NW, Suite 601
Washington, DC 20036
202.293.2450
aacte.org

American Association of Engineering Societies (AAES)
1801 Alexander Bell Drive
Reston, VA 20191
202.296.2237
orders@aaes.org
www.aaes.org

American Association of University Professors (AAUP)
1133 Nineteenth Street NW, Suite 200
Washington, DC 20036
202.737.5900
communications@aaup.org
www.aaup.org

American Board of Industrial Hygiene (ABIH)
6005 W St Joseph Highway, Suite 300
Lansing, MI 48917
517.321.2638
www.abih.org

The American Ceramic Society (ACerS)
550 Polaris Parkway, Suite 510
Westerville, OH 43082
866.721.3322
ceramics.org

American Council of Engineering Companies (ACEC)
1015 15th Street NW, 8th Floor
Washington, DC 20005-2605
202.347.7474
acec@acec.org
www.acec.org

American Design Drafting Association (ADDA)
105 East Main Street
Newbern, TN 38059
731.627.0802
www.adda.org

American Educational Research Association (AERA)
1430 K Street NW, Suite 1200
Washington, DC 20005
202.238.3200
www.aera.net

American Industrial Hygiene Association (AIHA)
3141 Fairview Park Drive, Suite 777
Falls Church, VA 22042
703.849.8888
infonet@aiha.org
www.aiha.org

American Institute for Medical and Biological Engineering (AIMBE)
1400 I Street NW, Suite 235
Washington, DC 20005
202.496.9660
info@aimbe.org
aimbe.org

American Institute of Architects (AIA)
1735 New York Avenue NW
Washington DC 20006-5292
www.aia.org

American Institute of Chemical Engineers (AIChE)
120 Wall Street, FL 23
New York, NY10005-4020
800.242.4363
www.aiche.org
Career Center:
646.495.1321
internships.aiche.org

The American Institute of Mining, Metallurgical, and Petroleum Engineers (AIME)
12999 East Adam Aircraft Circle
Englewood, CO 80112
303.948.4255
council@craftcouncil.org
www.aimehq.org

American Nuclear Society (ANS)
555 North Kensington Avenue
La Grange Park, IL 60526
800.323.3044
askanything@ans.org
www.ans.org

American Society for Engineering Education (ASEE)
1818 N Street NW, Suite 600
Washington, DC 20036
202.331.3500
www.asee.org

American Society for Engineering Management (ASEM)
200 Sparkman Drive, Suite 2
Huntsville, AB 35805
256.503.8482
asem-hq@asem.org
www.asem.org

American Society of Agricultural and Biological Engineers (ASABE)
2950 Niles Road
St. Joseph, MI 49085
269.429.0300
www.asabe.org

American Society of Civil Engineers (ASCE)
1801 Alexander Bell Drive
Reston, VA 20191
800.548.2723
customercare@asce.org
www.asce.org

The American Society of Heating, Refrigerating and Air-Conditioning Engineers (ASHRAE)
180 Technology Parkway NW
Peachtree Corners, GA 30092
404.636.8400
www.ashrae.org

American Society of Home Inspectors (ASHI)
932 Lee Street, Suite 101
Des Plaines, IL 60016-6594
847.759.2820
www.homeinspector.org

American Society of Mechanical Engineers (ASME)
Two Park Avenue
New York, NY 10016-5990
800.843.2763
CustomerCare@asme.org
www.asme.org

American Society of Naval Engineers (ASNE)
1423 Powhatan Street, Suite 1
Alexandria, VA 22314
703.836.6727
asnehq@navalengineers.org
www.navalengineers.org

American Society of Safety Professionals (ASSP)
520 N. Northwest Highway
Park Ridge, IL 60068
847.699.2929
customerservice@assp.org
www.assp.org

ASM International
9639 Kinsman Road
Materials Park, OH 44073-0002
440.338.5151
memberservicecenter@asminternational.org
www.asminternational.org

Association for Computing Machinery (ACM)
1601 Broadway, 10th Floor
New York, NY 10019-7440
212.869.7440
www.acm.org

Association for Environmental Studies and Sciences (AESS)
PO Box 7314
Halethorpe MD 21227
202.503.4638
support@aessonline.org
www.aessonline.org

Association for Women in Computing (AWC)
PO Box 2768
Oakland, CA 94602
info@awc-hq.org
www.awc-hq.or

Association of American Colleges and Universities (AACU)
1818 R Street NW
Washington, DC 20009
202.387.3760
information@aacu.org
www.aacu.org

Association of American Educators (AAE)
25909 Pala Place, Suite 330
Mission Viejo, CA 92691
800.704.7799
www.aaeteachers.org

Association of College and University Educators (ACUE)
745 Fifth Avenue, Suite 500
New York, NY 10151
866.628.2283
acue.org

Association of Collegiate Schools of Architecture (ACSA)
1735 New York Avenue
Washington, DC 20006
202.785.2324
info@acsa-arch.org
www.acsa-arch.org

Association of Construction Inspectors (ACI)
P.O. Box 879
Palm Springs, CA 92263
877.743.6806
support@assoc-hdqts.org
www.aci-assoc.org

Association of Environmental Engineering & Science Professors (AEESP)
1211 Connecticut Avenue NW, Suite 650
Washington, DC 20036
202.640.6591
bschorr@aeesp.org
www.aeesp.org

Association of Teacher Educators (ATE)
11350 Random Hills Road, Suite 800, PMB 6
Fairfax, VA 22030
703.659.1708
info@ate1.org
www.ate1.org

The Association of Technology, Management, and Applied Engineering (ATMAE)
486 Cornell Road
Blairsville, PA 15717
919.635.8335
www.atmae.org

Audio Engineering Society (AES)
132 East 43rd Street, Suite 405
New York, NY 10017
212.661.8528
www.aes.org

Automation Federation
P.O. Box 12277
67 T.W. Alexander Drive
Research Triangle Park, NC 27709
919.549.8411
info@automationfederation.org
www.automationfederation.org

Biomedical Engineering Society (BMES)
8201 Corporate Drive, Suite 1125
Landover, MD 20785
membership@bmes.org
301.459.1999
www.bmes.org

Board of Certified Safety Professions (BCSP)
8645 Guion Road
Indianapolis, IN 46268
317.593.4800
www.bcsp.org

Computer Research Association (CRA)
1828 L Street
Washington, DC 20036-4632
202.234.2111
cra.org

Council of Graduate Schools (CGS)
One Dupont Circle NW, Suite 230
Washington, DC 20036
202.223.3791
general_inquiries@cgs.nche.edu
cgsnet.org

Dictionary of Occupational Titles (DOT)
occupationalinfo.org

Engineering Education Service Center (EESC)
1411 Old Hardman Btms Road
Clarkesville, GA 30523
706.499.5011
www.engineeringedu.com

Engineers for a Sustainable World (ESW)
424 Sapphire Way
Pittsburgh, PA 15224
412.475.9531
info@eswusa.org
www.eswglobal.org

Engineers Without Borders USA
1031 33rd Street, Suite 210
Denver, CO 80205
303.772.2723
info@ewb-usa.org
www.ewb-usa.org

Environmental and Engineering Geophysical Society (EEGS)
1391 Speer Boulevard, Suite 450
Denver, CO 80204
303.531.7517
www.eegs.org

ETA International
5 Depot Street
Greencastle, IN 46135
800.288.3824
eta@etai.org
www.etai.org

Future Farmers of America (FFA)
6060 FFA Drive
Indianapolis, IN 46278-1370
888.332.2668
www.ffa.org

Health Physics Society (HPS)
950 Herndon Parkway, Suite 450
Herndon, VA 20170
703.790.1745
hps@burkinc.com
hps.org

Institute of Biological Engineering (IBE)
446 East High Street, Suite 10
Lexington, KY 40507
859.977.7450
info@ibe.org
www.ibe.org

Institute of Electrical and Electronics Engineers (IEEE)
3 Park Avenue, 17th Floor
New York, NY 10016-5997
800.678.4333
contactcenter@ieee.org
www.ieee.org

IEEE Aerospace and Electronic Systems Society
admin@ieee-aess.org
www.ieee-aess.org

IEEE Computer Society
2001 L Street N.W., Suite 700
Washington, DC 20036-4928
202.371.0101
help@computer.org
www.computer.org

IEEE Engineering in Medicine and Biology Society (EMBS)
445 Hoes Lane
Piscataway, NJ 08854
732.981.3433
www.embs.org

Institute of Industrial & Systems Engineers (IISE)
3577 Parkway Lane, Suite 200
Norcross, GA 30092
800.494.0460
cs@iise.org
www.iise.org

International Association of Certified Home Inspectors (InterNACHI)
1750 30th Street, Suite 301
Boulder, CO 80301
888.422.7233
fastreply@internachi.org
internachi.org

International Association of Electrical Inspectors (IAEI)
901 Waterfall Way, Suite 602
Richardson, TX 75080
800.786.4234
www.iaei.org

International Association of Machinists and Aerospace Workers (IAMAW)
9000 Machinists Place
Upper Marlboro, MD 20772
301.967.4500
www.goiam.org

International Association of Plumbing and Mechanical Officials (IAPMO)
4755 E. Philadelphia Street
Ontario, CA 91761
909.472.4100
www.iapmo.org

International City/County Management (ICMA)
777 North Capitol Street, NE, Suite 500
Washington DC 20002-4201
800.745.8780
icma.org

International Code Council (ICC)
500 New Jersey Avenue, NW, 6th Floor
Washington, DC 20001
888.422.7233
www.iccsafe.org

International Society of Automation (ISA)
67 T.W. Alexander Drive
P.O. Box 12277
Research Triangle Park, NC 27709
919.549.8411
www.isa.org

Manufacturers' Agents National Association (MANA)
6321 W Dempster Street, Suite 110
Morton Grove, IL 60053
877.626.2776
mana@manaonline.org
www.manaonline.org

Manufacturers' Representatives Educational Research Foundation (MRERF)
1301 Arapahoe Street, Suite 105
Golden, CO 80401
303.463.1801
certify@mrerf.org
mrerf.org

Marine Engineers' Beneficial Association (MEBA)
444 North Capitol Street NW, Suite 800
Washington, DC 20001
202.638.5355
mebahq@mebaunion.org
www.mebaunion.org

Maritime Administration of the U.S. Department of Transportation
1200 New Jersey Avenue, SE
Washington, DC 20590
202.493.0029
careersafloat@dot.gov
www.maritime.dot.gov

The Minerals, Metals & Materials Society (TMS)
5700 Corporate Drive, Suite 750
Pittsburgh, PA 15237
800.759.4867
tmsgeneral@tms.org
www.tms.org

Materials Research Society (MRS)
506 Keystone Drive
Warrendale, PA 15086-7537
724.779.3003
www.mrs.org

National Action Council for Minorities in Engineering (NACME)
One North Broadway, Suite 601
White Plains, NY 10601
212.691.1051
info@nacme.org
www.nacme.org

National Association of Biology Teachers (NABT)
P.O. Box 3363
Warrenton, VA 20188
703.264.9696
office@nabt.org
nabt.org

National Association of Elevator Safety Authorities International (NAESAI)
P.O. Box 4367
Mankato, MN
360.292.4968
naesai.org

National Association of Power Engineers (NAPE)
1 Springfield Street, Suite 1
Chicopee, MA 01013
413.592.9273
www.powerengineers.com

National Council of Examiners for Engineering and Surveying (NCEES)
200 Verdae Boulevard
Greenville, SC 29607
800.250.3196
www.ncees.org

National Fire Protection Association (NFPA)
1 Batterymarch Park
Quincy, MA 02169-7471
800.344.3555
www.nfpa.org

National Institute for Certification in Engineering Technologies (NICET)
1420 King Street
Alexandria, VA 22314
888.476.4238
socoffice@soc.org
www.nicet.org

National Institute of Food and Agriculture (NIFA)
United States Department of Agriculture
1400 Independence Avenue, SW, MS 2201
Washington, DC 20250-2201
nifa.usda.gov

The National Society of Black Engineers (NSBE)
205 Daingerfield Road
Alexandria, VA 22314
703.549.2207
adaa@artdealers.org
www.nsbe.org

National Society of Professional Engineers (NSPE)
1420 King Street
Alexandria, VA 22314
888.285.6773
www.nspe.org

Nuclear Energy Institute (NEI)
1201 F Street NW, Suite 1100
Washington, DC 20004-1218
202.739.8000
www.nei.org

Occupational Safety and Health Administration (OSHA)
200 Constitution Avenue NW
Washington, DC 20210
800.321.6742
www.osha.gov

O*NET OnLine
onet@onetcenter.org
www.onetonline.org
My Next Move:
www.mynextmove.org

Pathways to Science
P.O. Box 607
Damariscotta, ME 04543
207.424.1185
contactus@ibparticipation.org
www.pathwaystoscience.org

Society for Mining, Metallurgy & Exploration (SME)
12999 E Adam Aircraft Circle
Englewood, CO 80112
720.574.1256
cs@smenet.org
www.smenet.org

Society of American Military Engineers (SAME)
1420 King Street, Suite 100
Alexandria, VA 22314
703.549.3800
www.same.org

Society of American Registered Architects (SARA)
14 E 38th Street, 11th Floor
New York, NY 10016
212.520.2901
www.sara-national.org

Society of Biological Engineers (SBE)
c/o American Institute of Chemical Engineers (AIChE)
120 Wall Street, FL 23
New York, NY 10005-4020
800.242.4363
www.aiche.org/sbe

Society of Manufacturing Engineers (SME)
1000 Town Center, Suite 1910
Southfield, MI 48075
313.425.3000
www.sme.org

Society of Hispanic Professional Engineers (SHPE)
13181 Crossroads Parkway North, Suite 220
City of Industry, CA 91746
www.shpe.org

Society of Nuclear Medicine & Molecular Imaging (SNMMI)
1850 Samuel Morse Drive
Reston, VA 20190
703.708.9000
www.snmmi.org

Society of Petroleum Engineers (SPE)
222 Palisades Creek Drive
Richardson, Texas 75080
972.952.9393
spedal@spe.org
www.spe.org

Society of Women Engineers (SWE)
130 East Randolph Street, Suite 4500
Chicago, IL 60601
1.312.596.5223
hq@swe.org
swe.org

SpaceTEC
379 Cheney Highway
Titusville, FL 32780
PMB 207
321.567.5193
www.spacetec.us

Technology Student Association (TSA)
1904 Association Drive
Reston, VA 20191-1540
703.860.9000
general@tsaweb.org
tsaweb.org

U.S. Bureau of Labor Statistics (BLS)
Postal Square Building
2 Massachusetts Avenue NE
Washington, DC 20212-0001
202.691.5200
www.bls.gov
Occupational Outlook Handbook:
www.bls.gov/ooh

U.S. Food and Drug Administration (FDA)
10903 New Hampshire Avenue
Silver Spring, MD 20993-0002
888.463.6332
www.fda.gov

U.S. Green Building Council (USGBC)
2101 L Street NW, Suite 500
Washington, DC 20037
800.795.1747
info@usgbc.org
www.usgbc.org

USAJOBS
c/o U.S. Office of Personnel Management
1900 E Street NW
Washington, DC 20415-1000
202.606.1800
www.usajobs.gov

INDEX

A

B

F

G

H

I

L

M

N

O

P

Q

R

S

T

U

W

The Careers In... Series

Careers in Artificial Intelligence
Careers in Biology
Careers in Building Construction
Careers in Business
Careers in Chemistry
Careers in Communications & Media
Careers in Education & Training
Careers in Engineering
Careers in Environment & Conservation
Careers in Financial Services
Careers in Forensic Science
Careers in Gaming
Careers in Green Energy
Careers in Healthcare
Careers in Hospitality & Tourism
Careers in Human Services
Careers in Information Technology
Careers in Law, Criminal Justice & Emergency Services
Careers in Manufacturing & Production
Careers in Nursing
Careers in Physics
Careers in Protective Services
Careers in Psychology & Behavioral Health
Careers in Public Administration
Careers in Sales, Insurance & Real Estate
Careers in Science & Engineering
Careers in Social Media
Careers in Sports & Fitness
Careers in Sports Medicine & Training
Careers in Technical Services & Equipment Repair
Careers in the Arts: Fine, Performing & Visual
Careers in the Automobile Industry
Careers in the Music Industry
Careers in Transportation
Careers in Writing & Editing
Careers Outdoors
Careers Overseas
Careers Working with Animals
Careers Working with Infants & Children
Careers: Paths to Entrepreneurship